2. FUZZY SYSTEMS

CYBERNETICS AND SYSTEMS SERIES

Editor in Chief

J. ROSE

Director General,
World Organisation of General Systems and Cybernetics

Already Published
1. Philosophical Foundations of Cybernetics
5. Economic Cybernetics
6. Cybernetics and Society

Other Titles in Preparation

3. Artificial Intelligence
4. Management Cybernetics
7. Models and Modelling Systems
8. Automation and Cybernetics
9. Medical Cybernetics
10. General System Theory
11. Computers and Cybernetics
12. Neurocybernetics

FUZZY SYSTEMS

C.V. NEGOITA
Faculty of Cybernetics and the Institute of Management, Bucharest, Romania

ABACUS PRESS

First published in 1981 by

ABACUS PRESS

Abacus House, Speldhurst Road, Tunbridge Wells, Kent TN4 OHU

British Library Cataloguing in Publication Data
Negoita, C.V.
Fuzzy systems.—(Cybernetics and systems series: 2).
1. Decision-making—Mathematical models
2. Set theory
I. Title II. Series
658.4′033 HD30.23
ISBN 0-85626-164-5

Typeset by Pearl Island Filmsetters in Hong Kong
and printed and bound by
Tien Wah Press (Pte.) Ltd in Singapore

SERIES PREFACE

The inter- and trans-disciplinary sciences of cybernetics and systems have made tremendous advances in the last two decades. Hundreds of books have been published dealing with various aspects of these sciences. In addition, a variety of specialist journals and voluminous post-conference reports have appeared, and learned societies, national and international, established. These substantial advances reflect the course of the Second Industrial Revolution, otherwise known as the Cybernetic Revolution.

In order to extend the readership from experts to the public at large and acquaint readers with up-to-date advances in these sciences which are rapidly achieving tremendous importance and which impinge upon many aspects of our life and society, it was considered essential to produce a series of concise and readable monographs, each concerned with one particular aspect. The twelve topics constituting the first series are as follows (in alphabetical order).

Artificial Intelligence, Automation and Cybernetics, Computers and Cybernetics, Cybernetics and Society, Economic Cybernetics, Fuzzy Systems, General System Theory, Management Cybernetics, Medical Cybernetics, Models and Modelling of Systems, Neurocybernetics, Philosophical Foundations of Cybernetics.

The authors are experts in their particular field and of great repute. The emphasis is on intelligible presentation without *excess* mathematics and abstract matter. It is hoped each monograph will become standard reading matter at academic institutions and also be of interest to the general public. In this age of enormous scientific advances and of uncertainty concerning the welfare of societies and the very future of mankind, it is vital to obtain a sound insight into the issues involved, to help us to understand the present and face the future with greater confidence.

J. Rose
Blackburn

PREFACE

Trying to apply mathematics to soft sciences, researchers are susceptible to trends of fashion. The system approach has shown their readiness to accept a new methodology. It seems to me that not only does an examination of this methodology cast light on the challenges currently facing the human systems, but it is also a case study of a topic that has come of age.

There can be little argument that the expertise of System Theory is potentially relevant to soft sciences. But the way in which this potential can be fruitfully realized is less obvious. This book contends that the first step in the process of applying System Theory to human systems must be the development of a paradigm. The basic assumption is that synthesis means conflict resolution. Structural stability questions motivate the introduction of new mathematical tools, and fuzzy systems theory shows in a very precise way how modern algebra can be used in the understanding of decision processes.

The book consists of a series of essays. It is consistently characterized by the experience of the author as head of research staff at the Institute of Management in Bucharest, and owes much of its present form to the comments and contributions of the participants in seminars, in which the first concern was to formulate the questions before obtaining the answers.

This book owes its existence in large measure to the editor, Dr J. Rose, who convinced me that whatever work has been done by us in this field should be published in the form of an unified and readable presentation.

C.V. Negoita
Bucharest

CONTENTS

Chapter 1

INTRODUCTION

This book represents an attempt to explore the relationship between fuzzy systems theory and homeostasis. It is hoped that it will provide the conceptual framework, the basic methods and the perspective essential for a better understanding of a highly important subject. It hardly needs to be said that in a relatively small book such as this a big subject cannot be treated in a fully authoritative way. But the fundamental difficulty is not really one of space. Rather, it is our still limited and imperfect understanding of the causes and consequences of the process that we call 'qualitative change'.

The central goal of General Systems Research seems to be an orderly way of handling the shifting, interacting individual events of existence. Therefore, truth is relative because empirical existence itself is in a process of change, and stability is to be found only in the process of interaction. Some systems show a phenomenon of robustness in that they are stable despite substantial alteration of their structure. That is, some aspects of the balance of the system are insensitive to quite radical changes in seemingly importance aspects. This is precisely the case of cognition and decision-making processes.

It is quite natural to think of change as a succession of states in which each holds the stage for a while before being following by its successor. Perhaps for this reason, methods from mechanics have been tried to explain the change. However, if we accept the dynamics as conflict resolution, then homeostasis also means movement toward equilibrium, but the theory of homeostasis cannot be more satisfactory than that offered by the framework of mechanics. *Autopoietic processes*, i.e. organic growth, require a better understanding of the concept called 'synthesis'. It seems to me that a good model for this concept could be formulated using the notion of 'pullback' introduced by the category theorists. Initial contradictions are reconciled in higher synthesis, for to remove differences is to achieve the purpose of equilibrium, and this process can be formalized, represented and modelled.

The development of General Systems Research at the moment is characterized by its own mathematical concepts, the penetration of mathematical methods into traditionally soft sciences now being generally acceptable. Along with the use of mathematics, there is taking place a saturation of mathematics itself with set-theoretical concepts, use of mathematics having become almost equivalent to using set theory. But the language of set theory is not always adequate for expressing the purposes of soft sciences, and applied mathematics is not merely an application of set theory to practical problems but something else. Fuzzy set theory is an example of mathematics based on a different concept.

One aspect of General Systems Research is its development in a direction that will enable more and more complex systems to be analysed or designed. The trend of fuzzy systems reflects the desire to represent human systems. The major unsolved problem in the field of General Systems Research seems to be that of transcendence. If we explore the mode of cognition as a procedure for describing complex systems and the way in which natural language is involved in this process, we see that mankind copes with complexity by assembling partial representations, and that learning means conceptualization, which is concerned with getting a feel for the whole. The unfolding of multi-criteria decisions follows the same procedure. Transcendence signifies a trend towards totality, completeness and wholeness. As such we are speaking of timeless and universal concepts or criteria that clarify everything and establish relationships. Transcendence seeks a unification of particular existence in such a way that every detail would be a unit of the totality. There is, indeed, a sense in which transcendence expresses the synthesis of empirical existence. If we accept that transcendence discloses a constant openness in which man grasps freedom, then it is high time to obtain a deep insight into this natural phenomenon.

1.1. FUZZY SYSTEMS

The expression 'fuzzy system' was coined by Lotfi Zadeh in 1965. It is generally taken to embrace the whole field of imprecisely described systems, but though this implies the use of verbal descriptions, there is a tendency to regard the subject as providing a substitute for methods such as this.

There is, however, no recognizable or accepted body of theory that can be considered as an established discipline. For this reason, those writers who have attempted to survey the field have studiously avoided the use of a title that would suggests any claim that they were dealing with a subject which has recognizable boundaries.

The problem of finding a way to handle inexact descriptions has created a challenge that is attracting the attention of workers from an increasing number of disciplines. The proliferation of different kinds of approach has tended to cloud the issue, and this book was born of a conviction that, within the limits of our present knowledge, we can isolate and exploit some principles.

Chapter 2 is an attempt to explain the need for a study of fuzzy systems and provides some bibliographical hints. The principles, which to my mind are fundamental to the problem, form the subject of Chapter 3. Chapter 4 and 5 give sketch the mathematical background, Chapter 6 introduces the concept of pullback as a model of synthesizing processes, and Chapter 7 illustrates the theory.

The principal theme of the book forms a continuing argument through Chapters 3 to 7 and comprises an integrated study of fuzzy relations viewed as states of dynamic systems.

I shall define fuzzy systems more comprehensively later, but I think that for the purpose of this introduction and for drawing the interested reader further into the book, I would describe them as systems of systems, or systems of logical objects. The object of studying such systems is to gain understanding so that they may be more readily controlled, just as systems in the natural sciences are.

Most of the techniques described will be illustrated by examples from my own work and that of my colleagues. This is not merely conceit, but based upon a belief that first-hand knowledge is best. The views presented here might have been elaborated with greater mathematical refinement. But there may be some virtue in articulating new ideas briefly and simply to an intelligent unprofessional audience whose response can give the enterprise an authentic air of intellectual adventure. The book will also provide the specialist with some original but worthwhile concepts and some new ways of looking at well-known problems.

In the course of the book, there will be found references to the writings of other workers in the field, when it has been found possible to interpret their work in relation to our model. These are scattered throughout the book with appropriate citations when they occur.

This book is not intended to be a treatise on the field, nor is it intended to be a critique of the work of other schools, nor a comparative evaluation of the success and failure of different approaches to the subject. It is intended to set forth in an orderly manner the motivations and viewpoints of a school of thought that cherishes the belief that the field of fuzzy systems can and should be discussed with mathematical skill.

1.2. SYMBOLS AND TERMS

Now, a short explanation of mathematical symbols and terms used in the book. Elements are indicated by lower case letters x, y, z. Sets are designated by capital letters A, B, C, etc. Set membership is indicated in symbols as $x \in A$ which reads 'x is a member of A' or 'x belongs to A'. Given two sets A and B the Cartesian product of A and B is the set of ordered pairs in which the first element belongs to A and the second to B, denoted as $A \times B = \{(x, y) | x \in A \text{ and } y \in B\}$. If A and B are two sets we write $A \subset B$ to indicate that A is a subset of B, i.e. that every element of A is also an element of B. $A \cap B$, called the intersection of A and B is the set $\{x | x \in A \text{ and } x \in B\}$. $A \cup B$, called the union of A and B, is the set $\{x | x \in A \text{ or } x \in B\}$.

The set of all corresponding pairs (x, y) such that y is the single value associated with the element $x \in A$ constitutes a function f on A. It is not necessary, however, that to each y there corresponds only one x. The inverse function f^{-1} is the set of correspondences between each $y \in B$ and the subset of A which is mapped into y by f. We use the notation $f: A \rightarrow B$ for 'f is a function from the set A to the set B', i.e. 'to each $x \in A$, f assigns an element $f(x) \in B$. The composition of two functions f, g—denoted by a small circle, $g \circ f$, means a succession of two arrows $A \xrightarrow{f} B \xrightarrow{g} C$.

A *lattice* is a partially ordered set such that all two elements have a greatest lower bound and a least upper bound. The existence of these unique bounds for each pair of elements in a set A implies the existence of two functions $A \times A \rightarrow A$, called *supremum* and denoted as $\vee$ and called *infimum* and denoted as $\wedge$.

A *category* is a formal system. Formal means explicit, not merely understood. A formal system has a collection of axioms from which we obtain theorems by the repeated application of a number of rules. Formal means also 'of the essence of the thing' rather than its matter. The Aristotelian logic uses categories as fundamental classes among which all things can be distributed. Categories viewed as formal systems are also fundamental classes among which mathematical things can be distributed.

At many stages of the development the discussion will tend to be somewhat abstract. The reader is invited to relax and enjoy the trip. The ideas are, in fact, simple.

Chapter 2

AN APPROACH TO COMPLEXITY

2.1. WHY FUZZY SYSTEMS?

The systems concepts have been under-utilized in relation to the obvious and urgent need for their employment in dealing with complex problems. This book is an attempt to answer the question: what can be done? The main theme is that although many scientists boldly apply existing models of System Theory to complex problems, little research is devoted to the foundations of this model building.

System Theory is defined as a body of concepts and techniques used to analyse and design systems regardless of their nature. The important aspect of systems analysis and design is the development of a model that describes the 'cause and effect' relationships between the system variables. But an exact description of any real system is virtually impossible. Our inability to make precise statements about complex behaviours is a fact we have to accept and adjust to. Complexity is associated with description rather than being thought of as an intrinsic property of objects. Hence, we may well consider reducing the complexity of an object, not by changing that object, but by changing our views about it.

People who have attempted to analyse complex problems from a systems point of view have frequently remarked that symbolic language describing a model is too poor to convey all the nuances a person can carry in the mind, which is why social scientists require the medium of ordinary, natural language and its richer content. In some ways it is unfortunate, though almost inevitable, that the modern concept of the systems approach has evolved out of technical problems burdened with the artificial language of pure mathematics. Many system theorists have failed in what they tried to do because they never looked at the language they used. If they had, they might have begun to question some of the bases of what they were doing.

Conventional systems approach has tended to assume that what needs to be done is to survey the whole system. The fuzzy systems approach challenges this assumption and proposes to construct a global representation on the basis of assumptions about partial representation. In this way we are able to hold different imperfect representations of reality in hand at the same time. A theory based on consideration of individual representations, such as the one proposed by fuzzy systems theory, is inherently richer in its consequences than a theory that treats a system as if it were an individual representation. A description depends not only on each partial representation but also on interaction among representations.

Humans obtain knowledge of their environment by perception, the link between the outer world and the inner world, between what is commonly called 'material' and what has been named 'mind'. We see, hear, smell, taste, feel heat and cold, and form size and texture. Finally, all the perceptions merge into a complex background and are no longer isolated. A complex system is perceived as a single totality. Abstraction is viewed as a set of interrelated mental images. Words are mere labels for general functions. The interesting question is not whether such functions exist, for they are clearly implied in the phenomenon we call mental experience, but how do they develop? How do they work?[1]

Each word of natural language is inherently vague. It is an essential characteristic of a vague concept that the boundaries of the domain of its applicability are not fixed and, therefore, we do not know precisely where this domain ends and some other begins; that vagueness is not thought of as inherent in the real world, but rather in our use of words. Put in another way, inexact concepts are associated with the inability to decide, which is generated by conflict. The principle of the excluded middle is regarded as inapplicable in these situations because the notion of contradiction cannot be directly applied there. People understand vague concepts as if these were represented by fuzzy sets. The assumption underlying fuzzy sets is that the transition from membership to non-membership is a gradual specifiable change. Membership in a fuzzy set is specified by a mapping. This mapping can be further used in fuzzy inferences. People manipulate vague concepts as if they are processed according to the rules of fuzzy logic; several studies appear to support this theoretical position.[2]

People face complexity because they have the ability to see essential features by a process of abstraction. Abstract concepts are amalgamations of subconcepts, each of which has different conflicting meanings. This is, indeed, a powerful model with many potentially fruitful applications and it deserves serious attention. Abstraction is the way people cope with complexity.

The moral is clear: fuzziness far from being a difficulty is often a convenience. The theory of fuzzy systems may well have a substantial impact on scientific methodology, particularly in the realm of soft sciences. Many people suppose that once you say a process is subjective, that it depends on human judgement and beliefs, analysis ends there. I disagree. Subjective processes are fully as amenable to scientific understanding as any other process. Some results in this field are encouraging, for these results tend to show a surprising and beautiful unity hidden in the apparent diversity of approaches. The key to the unification lies in conflict resolution. As I have already said, inexact concepts are associated with the inability to decide, which is generated by conflict. Domains of applicability are always overlapping. Our work makes the fuzzy set approach to hierarchical systems completely explicit and extends their applicability by demonstrating that the ability to coordinate means transformation of fuzzy sets. The existence of a conflict at one level generates the next level. This movement from one level to the other is precisely the dynamics of a fuzzy system. The result of this movement is compact representation, i.e. abstraction.

This is exactly the sense in which one distinguishes between an organic and a mechanical theory of movement. The growth of a fuzzy system is an autopoietic model; a spontaneous order, corresponding to an organism, is achieved. The structure is not designed from without, the forces that determine such an order are inner conflicts.

Conflict resolution at a higher level means composition of fuzzy sets. Under such a composition, a new fuzzy set is generated. Fuzzy sets are associated with ordering in a particular aspect. I define it by referring to an ideal point. Therefore, by composing fuzzy sets, the autopoietic process moves the ideal point. In other words, the dynamics of fuzzy sets means dynamics of ideal points. Self-coordination induces a displacement of ideal and can model the process of dynamic decision-making. In technical language, the absolute ideal corresponds to the last level in the hierarchy of fuzzy sets, and, in the jargon of category theory this is a terminal object in the category of fuzzy sets. Moving up in the hierarchy means to become aware of new criteria or points of view. Scientific knowledge represents partial truth and there may be material changes as new knowledge accumulates. Clearly, abstraction means unification.

There are many problems in management science in which hierarchically organized but incompletely specified information is important. Planning is notorious for its often ambiguous underlying assumption and its nearly always, vague, multilevel approach. Take, for instance, the linear programming problem. When speaking of objectives or constraints it is most often understood that these are well-known. Real life problems frequently lack this property. In fact, management decision-making involves the choice of an action from alternative courses of action to achieve fuzzy objectives. The significance of studying optimization problems in fuzzy environment has been recognized and some problems have been solved.

A fuzzy programming problem is one of multicriteria optimization. Constraints and objectives are fuzzy sets, i.e. functions; there is no distinction between means and ends. They are functions defined on the same carrier and we are speaking about their confluence. To solve a fuzzy programming problem means to move the decision to a higher level, to consider all the points of view as a whole. In the language of hierarchical systems theory this is coordination. To coordinate is to find a special kind of function that describes the relationship between the global objective and the local ones. This function must preserve order, a condition named *monotonicity*, and the intersection of fuzzy sets is clearly a monotone function.

Critiques of the systems approach remark that in all systems analysis, there is a moral overtone which can also be seen in planning and other activities where intellectuals become involved in trying to improve society; it addresses itself to collective properties and is incapable of considering people as unique individuals. If one looks at the linear programming model, one will see that, say the critics, the model for a machine would be the same as a model for a social system. Systems analysts are do-gooders; they are trying to help to improve social systems. But in all their studies they treat individuals as means rather than as ends. Hence, in some sense, they fail to deal with the moral dimension. We need a different approach if we are going to be honest. Essentially, what must change is the level. If different individuals prefer different alternatives there is a conflict again. As in any conflict situation, it can be solved by considering a move towards a higher level. We can bring the merits of judgement and beliefs to the linear programming model in a quantitative manner by using fuzzy sets instead of numbers as parameters; this is a compact representation.

Difficulties with the present-day soft sciences do not include lack of good results, but *do* include lack of theoretical foundation. It is hoped that the fuzzy systems theory will put us in a better position to tackle some of the questions concerning organic growth by understanding the rules of autopoietic processes.

Modelling in an abstract setting requires, and when done automatically provides, a refined analysis of the fundamental concepts and constructions involved. Thus, a foundation study of organisms is an inevitable and welcome byproduct. For example, what is called displacement of preferences in decision-making, and what is called displacement of possibilities in economics, is more exactly analysed as dynamics of fuzzy states, described here as transformation of fuzzy sets, i.e. arrows in the category of fuzzy sets.

The dynamics of fuzzy systems is a deterministic movement of imprecise evaluations. The idea of using possibility as an uncertainty variable was considered in order to lay down precise lines of action, and to count upon their being followed. Since we need at least to know the circumstances that will prevail, including the actions that at future times will be taken by other people not subjected to the planned instructions, policy must legislate for tolerances. In order that policy may command loyal and resolute endeavour it must offer some assumptions of being able to cope with situations that cannot be foreseen in detail; this is robustness.

The mathematical System Theory extensively promoted during the 1960s has come under a barrage of criticism, and a new movement—back to the basics—has come to the fore. There are complaints that mathematical System Theory produced computational cripples who are seriously hindered in their attempts to use mathematics in soft sciences.[3] The past decades have served to demonstrate that System Theory and soft sciences move in separate orbits. Those orbits will have to coalesce if major reforms are ever to be successfully implemented. The development of a fuzzy systems theory is posed as a challenge for both applied mathematics and soft sciences.

The struggle to reconcile mathematics and the soft sciences has its parallel in automatic control, also dominated by pure mathematics. Among control theorists there have emerged, also, advocates of the need to develop a logic of inexact concepts.[4] Taking a retrospective view of the development of the theory of automatic control from its beginning to the time when interest in variational techniques reached its peak, they observed that this theory is permeated with unfulfilled hopes. Many problems remain unsolved to this day. The problems in question reduce, in the main, to the analysis of an equation which depends on parameters. In general, simple solution properties abide in complex, difficult to describe and highly irregular, parameter regions. By questioning the language used to describe these regions, they observed that applied mathematics is not a second-class kind of mathematics, nor is it merely an application of pure mathematics to practical problems, but a different kind of discipline which should be based on a different concept of precision.

Fuzzy set technology has already been born and is growing. It is based upon subjective judgements and objective aggregation rules are required to put them together. Treating autopoietic processes as fuzzy systems has made possible recognition of equal structures from which can be developed fundamental laws and principles that operate commonly at different levels of organization. Its adoption as a scientific approach does much to mitigate the mathematics-nonmathematics dichotomy and improve communication across scientific disciplines.

2.2. THE MILESTONES

A few key references make it fairly easy to penetrate the literature of fuzzy sets and systems rapidly and to a reasonable depth. Almost all the papers in this field are listed in the huge bibliography, published in 1977 by Gaines and Kohout,[5] which consists of some 1150 items, each keyword-indexed, about 750 of them classified as being related to fuzzy systems theory and its applications. The remaining items deal with closely related topics in many-valued logic, linguistics, the philosophy of vagueness, and similar topics. These background references are annotated in an initial section that outlines the relationship of fuzzy systems theory to other developments and provides pointers to various possible, fruitful interrelationships. As stated by the authors, the organization of this preamble—the best study I know—is aimed primarily at helping those interested to find their way around the bibliography, and secondly as a tribute to Lotfi Zadeh, who opened this area of study and has been a consistent driving force behind its development.

The basic aspects of the theory can be rapidly gleaned from two papers of Zadeh,[6,7] entitled *Fuzzy Sets*, the second one being written after an interval of ten years. An extensive presentation of the basic concepts is given by Kaufmann.[8] His textbook provides a concise, rigorous and up-to-date introduction to the theory of fuzzy sets and is intended to prepare those with little or no previous knowledge of this field for work with advanced and specialized texts. Although written in an applied mathematical style, it does not demand an advanced mathematical background and is illustrated by many examples and exercises. A broad spectrum of applications of the theory of fuzzy sets, ranging from its mathematical aspects to applications in human cognition, communication, decision-making and engineering system analysis are presented in a volume containing the papers contributed by the participants in the US-Japan Seminar on 'Fuzzy Sets and their Applications', held at the University of California, Berkeley, July 1974.[9]

A compact presentation of the algebraic theory of fuzzy sets can be found in a book written by Ralescu and me and published in 1975.[10] Here we have introduced the so-called theory of representation (the arrow $X \to (0, 1)$ was reversed as $(0, 1) \to p(X)$, whose significance was not seen until recently. I refer to the attempt by Boyd[11] to present topoi as models of fuzzy sets.

In a long and serious essay, Arbib[12] reviews these three books and notices that the most disturbing feature of the literature collected within these volumes is its insularity. The applied papers, he says, use the language of fuzzy sets to tackle some problems, but never provide a hard-headed comparative study to show that the method is better than any other. However, very respectable international journals like *Man-Machine Studies* or *Kybernetes* devoted special issues to fuzzy topics, and meetings such as I.E.E.E., ORSA/TIMS, and S.G.S.R. allocated special sessions to the topic.[13,14,15] The international journal *Fuzzy Sets and Systems* published by North-Holland, Amsterdam, seeks to present papers of real significance and broad interest (1978). Its history can be traced to the discussions of the second meeting of the European Working Group on Fuzzy Sets.[16]

I shall now mention a few references that mark important milestones in the evolution of the theory of fuzzy systems. One, due to Zadeh,[17] is an outline of a new approach to the analysis of complex systems and decision processes, a substantial departure from the quantitative techniques of systems analysis. The linguistic variable is suggested in place of, or in addition to, numerical variables. Simple relations between variables are characterized by fuzzy conditional statements. By relying on the use of linguistic variables the approach provides an effective means of describing the behaviour of systems too complex to admit of precise mathematical analysis. This is the case when one tries to explain the mechanisms behind social or psychological phenomena. In soft sciences it is customary to phrase the current knowledge in natural language. The drawback with regard to natural knowledge is that it does not provide an investigator with a means of exploring systematically the implication of the stated knowledge. If the knowledge consists of stated causal relations between time-varying variables, there is no way it could be simulated. In such situations, one resorts to rephrasing the model in mathematical language. Because of its precision the model will be only one among a number of possible models. What Zadeh says is that it might be possible to combine the virtues of the two languages by designing an auxiliary language that resembles natural language with respect to vocabulary, syntax, and semantics and where meaning is represented mathematically as fuzzy sets. Wenstøp[18–20] calls it 'deductive quasinatural language'.

A collection of implied statements (if A then B), which causally link input and output fuzzy sets can replace the usual mathematical model of control theory, and the composition operation allows the model to be used in the same way that matrix multiplication allows a state space model to be used. The system $x_{k+1} = Ax_k$ where x, u are fuzzy sets and A is a fuzzy relation, has been lately extensively studied by Mamdani,[21] Kickert[22] and Tong.[23] The conditions in which a fuzzy model may be appropriate are discussed and a standard notation is adopted to help focus attention on the main features of such a model.[24] Ways in which fuzzy models can be constructed are considered together with some difficulties that can arise.[25–27] The term 'fuzzy model' is taken to mean a finite set of fuzzy implications or rules which together form an algorithm for determining the outputs of a process from some finite number of past inputs and past outputs. The algorithm is often conveniently represented by a fuzzy transposition table.[28,29]

Studying this kind of system, Sanchez[30] solved the problem of invariants in therapeutic recommendations. Given a fuzzy relation A between medical symptoms expressing the action of a drug on patients in a given therapy, one can determine the greatest intensity of each symptom on which A produces no effect. When x_{k+1} equals x_k, as above, one says that x_k is an *eigen fuzzy set* associated with the given relation A. In a recent analysis of the dynamic behaviour of a linguistic model, Kickert[31] showed that the most important issue is the problem of convergency of the model to some stable state. This convergency problem is treated from the point of view of the fuzzy relation matrix. He presents some results concerning the powers of a fuzzy matrix, using theorems given by Sanchez at the solution of fuzzy relational equations.[32]

The building of such models takes place in a trend in which the modelling achievements are considered from a very pragmatic viewpoint. One attempts to embody the experience of experts in the framework of a verbal model. Running parallel with these ideas, it is possible to observe a few attempts to define languages aimed at simplifying the analyst's task in writing the models. Examples can be found in the papers of Lefaivre,[33] Zadeh[34] and Adamo[35,36]. Another milestone in the evolution of the theory of fuzzy systems is a paper written by Bellman and Zadeh[37] on decision-making in a fuzzy environment. Here, constraints and objective functions are all fuzzy sets. In such situations the relationships between constraints and objective functions are fully symmetric, i.e. there is no longer a difference between the former and the latter. This paper generated a chain of literature related to the fuzzy programming problem.[38–48]

The idea of representing a fuzzy set as a family of sets, first published in *Kybernetes*,[49] leads to the so-called problem of robust programming.[50–53] It is now possible to use fuzzy sets instead of numbers in linear programming problems. Recently, Dubois and Prade[54] have shown that the resolution of systems of linear fuzzy constraints is made easier by the use of fuzzy numbers analytically represented through a given type of membership function and three parameters. Although the fuzzy character of the problem is lost, the results are of practical value. More details about robust programming can be found in the journal *Economic Computation and Economic Cybernetic Studies and Research*, published by the Faculty of Cybernetics in Bucharest.[55]

A categorical approach of the representation theorem[56] was reported at the Third Congress of Cybernetics and Systems held in Bucharest in 1975, and the attempt to embody this theorem in the theory of elementary topoi was published in 1978.[57] The idea of *C*-set, due to Flondor[58] permitted a representation theorem due to Ralescu.[59]

We shall now turn our attention to the last example of synthesis, namely the retrieval process. The relationship between understanding knowledge and information is presented at large by Kochen.[60] He states that 'knowing' means question-answering, and 'understanding' means question-asking. Anyway, the retrieval process is considered as a matching process between a request specification and a set of information items. The general approach with most retrieval systems has been to make the matching by attaching to each information item key words, or descriptor, that indicate content. One feature of the user-friendly systems is the capability of tolerating fuzziness in descriptors. Schek[61] shows how the word context can be used to define similarity measures that model the intuitive human-like notion of similarity. In fact, information retrieval systems are too complex to be susceptible of exact analysis. Fuzziness is the inherent characteristic of such systems.[62] Here decisions are fuzzy in nature, i.e. decisions concerning the relevancy of an information item to a query, or the usefulness of a term as descriptor. Tahani[63] and Radecki[64] give models that take account of these realities.

An information retrieval system is one that compares the specifications of required items with stored items, and lists all the items in some defined way. The main criteria of success of such systems is the extent to which the utilized definition of matching, or the formulated query, yield items judged by users of the system to be relevant to their intentions. If the answer of the system is a fuzzy set, then to use many matching functions, or many queries, one must synthesize a unique answer. This synthesis is discussed in two of my works[65, 66] published in 1970. The idea of representing synthesis as a pullback was presented in some recent papers[67–69] and stems from those 1970 papers.

Chapter 3

CONCEPTS, DECISIONS AND VAGUENESS

While the desire to explain the theory of fuzzy systems was the principal motivation of this book, it is impossible to develop a theory without basing it upon reality. I found it essential to start with a fairly typical representation of how cognition and decision-making come about, and then to devise a theory that is not only consistent with the observations, but one that can also make predictions. I feel confident that the dynamic theory of this Chapter 3 provides a better understanding of synthesis processes, which are to be presented in Chapter 6, which is really an extension of this chapter, and has been written with both the non-specialist and specialist of System Theory in mind.

In something as important as the soft sciences, anything that even begins to free us from empiricism or to set empiricism free for new advances makes a contribution. This chapter—I hope—is such a contribution.

3.1. CONCEPTS

A fundamental element in all intellectual knowledge, and a condition of all rational communication, is the concept. Concepts are means of economy in human cognition. Beneath the multiplicity and variety of real objects (objects thrown before the mind), we can penetrate the unvarying pattern, the logical object, which is realized equally by all the individual real objects. We say 'tree' for a large number of individual trees and for different trees. Besides seeing the differences between things, we see uniformities, patterns of likeness running through them. It is, in fact, only through these invariable patterns of sameness that we can say, in the strict sense, that we know (see) a thing. The word form (*eidos*) is rooted in the Greek verb *idein* (to see). Intellect is rooted in the Latin *intus*, and *legere, to read within,* means to discern meaningful patterns within the manifold of experience. The material objects of knowledge are thus the real things that we know, and the logical objects of knowledge are those intelligible aspects or forms through which we come to know them. What we know is this tree (material object) and we come to know it through its stable logical object 'tree'.

We assert the independent existence of ontological objects distinct from the epistemological objects by which the former are known. Images are not the objects they refer to. Objects exist independently of concepts, and concepts have meanings only because they themselves endow them with meanings. In other words, we say that platonism and nominalism are indefensible extreme positions. What we seek to do is to characterize objects. We discover character with individuals. In conceiving individuals we may be led to use the same term for them. Then we have a class and speak of common connotation. We are also likely to speak of the members of a class as similar, similar with respect to some characteristics. After all, we are thinking about objects, not guessing at them.

We say 'tree' but we cannot have a representation of it. The concept cannot be pictured or imagined in any way. There are many representations, an infinity, but there exists no unique representation. It is said that the vogue of concept as a word that will dignify any idea from a passing fancy to a fully-fledged plan has probably blurred forever a useful distinction between them. But those who like to call things by their right names continue to mark a difference between 'concept' (the abstract general notion of any entity) and conception (the particular image, shape or set of features with which the empty concept may be filled). Thus, the concept God is the idea of a supreme ruler of the universe. All the religions propose different conceptions. For any nameable thing there can be only one concept, but the conceptions are many. In some sense, every individual cosmology is unique. In other respects all individuals share common beliefs about reality. However, no cosmology is reality; it is a human perception of reality. It is like a map of a territory that has been only partly explored and perhaps never will be completely. The traveller must therefore rely to some extent on his own wits, using his map but remembering always that it is an imperfect representation of reality.

'Tree' is something that expresses a possibility, a certain range or extent of reality. Possibilities are opinions. Then the inference rule is a formal rule that specifies how those opinions should be revised or modified in the light of new information. And the revision of opinion in the light of new information, otherwise called information processing, modelling and the like, is one of the most important of human intellectual activities.

Concepts are arranged in a sort of hierarchical structures. Up to the top is the 'good life', or something like that, and way down is the fact that we prefer steak to lobster for dinner because we might not sleep well if we ate lobster. The general principle is: do not work with too detailed and specific sets of dimensions, or we shall have too many dimensions and none of them will be very important.

All concepts are vague. To say that a concept is vague is to say that there are occasions when there is no definite answer to whether or not it applies to something. Our inability to say whether this object is a tree or a bush is not the result of lack of information about such things as height or colour. No additional information would settle the matter, except indirectly by leading us to tighten up the meaning of the word. The indeterminancy is due to an aspect of the meaning of the term rather than to the current state of knowledge.

If one surveys varieties of trees one finds that the trees resemble each other in a variety of ways without there being any one resemblance that exists between every pair. Properties or qualities give room for ordering. In a sense, to order means to arrange, and all orders are as arbitrary as are arrangements. But when we say that we are ordering real objects, we are really looking for certain relations between them, which themselves generate a certain arrangement. Ordering, therefore is an operation performed on objects when we discover a relationship that exists among its members.

The Cantorian set theory regards a set as given if, for each object, it is possible to draw a conclusion that it is or is not an element of the set. The philosophical meaning of this definition is that a set is understood to be a multiplicity of things that can be thought of as one. We are presupposing our ability to make a single formation out of many objects. When we speak of elements of a set we perform a certain operation of identification beforehand.

On the other hand, speaking about concepts, what we really do is to say that a predicate does or does not hold, although we cannot identify the objects. The philosophical meaning of the concept is unity thought of as many. Logical instruments have a particular ontological status.

Literally, to be universal means to be one with respect to many (*unum versus alia*); and interpreted in a logical context, this means that concepts, while in a sense meaning one thing, nevertheless are applicable to many things.

There now arises the question of how a concept can be represented.

3.2. SETS AND FUZZY SETS

'Tree' and 'bush' are vague concepts because there are occasions when their application is not definite. This indeterminancy does not mean that one lacks a sufficient knowledge of the concept or some facts, but the concept itself is such that its relation with the state of affairs in question is uncertain.

Let us see what is happening when vague concepts are represented as sets. I shall follow the reasoning given by Goguen[70]. Consider the set of trees, T. Then, $t \in T$ means 't is a tree'. Let B be the subset $B \subset T$ of all bushes, and let $h(t)$ be the height of $t \in T$. Then, h is a function from T to R, the set of real numbers. In the so-called functional notation this is written $h: T \rightarrow R$, and $h(t_1) \geqslant h(t_2)$ means t_2 is at least as small as t_1. Suppose that there is at least one bush, i.e. $B \neq \phi$, where ϕ is the empty set. Now, consider the inductive premise: if we call t_1 bush and we cannot discern the difference in the heights of t_1 and t_2 then we would also call t_2 a bush. Finally, we need the truism that a tree smaller than a bush is a bush. It follows from a finite iteration of *modus ponens* that any given tree being of finite height is a bush. Starting with a tree, call it t_1, known to be a bush, we choose a tree t_2, α centimetres taller than t_1, another t_3 β centimetres taller than t_2, and so on. The inductive premise is satisfied for t_1 and t_2, so t_2 is a bush. Therefore we can apply it to t_2 and t_3 to conclude that t_3 is a bush, and so on, until we conclude that any given tree is a bush. We have reached a conclusion contrary to our intuition from premises which did not appear unreasonable. That there are bushes, and that a tree smaller than a bush is a bush, seem to be two assumptions that are beyond question. So we are left with the inductive premise. It seems that we must abandon the idea that the law of excluded middle applies to such vague concepts as bush, for as soon as we agree to represent a bush by a subset B of trees, we impose an exact boundary between bushes and trees.

Representing vague concepts by sets and deduction by methods of traditional logic does not yield an adequate model of our customary use of vague concepts, for it has been shown that this representation leads to paradoxical conclusions. What actually happens is that we become less and less convinced of the smallness of trees as their height gradually increases.

A different representation of the concept 'bush' avoids the paradox. The main idea is to represent it by a fuzzy set. Let [0, 1] be the unit interval, the set of all real numbers between and including zero and one, and a function $B: T \to [0, 1]$. First of all, B should be a function only of the height of trees, that is $B(t) = f(h(t))$ for some function f, where h is the height function as above. f should be continuous and monotone decreasing, for we judge a taller tree as being less small and our judgement seems to change smoothly with height. Furthermore, if there are bushes, $f(h)$ should equal 1 for some value of h and if high trees are to be definitely not 'bush', $f(h)$ should approach zero as h gets large. These conditions correspond to assumptions that a tree smaller than a bush is a bush and that there are both bushes and trees. Many arguments about fuzzy sets do not depend on particular values of functions, but only on such general properties.

Note that we are not concerned with the likelihood that a tree is a bush after many trials; we are concerned with only one observation.

We can also define fuzzy relations, for which the degree of relationship can vary continuously between zero and one. If X and Y are sets, $X \times Y$ denotes the set $\{(x, y) | x \in X, y \in Y\}$ of all ordered pairs (x, y) of elements from X and Y. A fuzzy relation is a function $R: X \times Y \to [0, 1]$.

We are now prepared to examine what becomes of the paradoxical reasoning when 'bush' is represented by a suitable fuzzy set instead of a subset. The inductive premise provided a way of deducing $t_2 \in B$ from $t_1 \in B$ for certain pairs of trees. The inductive premise can be reformulated as the assertion $t_1 \in B \to t_2 \in B$, where $\to$ is read as 'implies'. This is a relation $R(t_1, t_2)$. We had $R(t_1, t_2) = 1$ for $(t_1, t_2) \in \{(t_1, t_2)/0 \leqslant h(t_2) - h(t_1) \leqslant \alpha\}$.

When B is a fuzzy set the inductive premise should provide a way of deducing the truth value $B(t_2)$ of 't_2 is bush' from the truth value $B(t_1)$ of 't_1 is bush' for appropriate pairs (t_1, t_2). The simplest approach is to let $R(t_1, t_2)$ be the fuzzy relation 'relative smallness of t_2 with respect to t_1' defined by

$$R(t_1, t_2) = \frac{B(t_2)}{B(t_1)}$$

We, therefore, have the basic deductive relationship

$$B(t_2) \geqslant R(t_1, t_2)\, B(t_1) \quad \text{for all} \quad (t_1, t_2)$$

If t_1 and t_2 are of similar stature, $R(t_1, t_2)$ is nearly one. The validity of the deductive process is measured by $R(t_1, t_2)$. The deductive process becomes less and less valid as the number of application of *modus ponens* increases.

Now suppose we have $t_1 \in B$ such that $B(t_1) = 1$ and some arbitrary t such that $h(t_2) \geqslant h(t_1)$. Then we may construct a finite sequence such that $h(t_{i+1}) \geqslant h(t_i)$.

What has been said can also be couched in philosophical terms. If we note that one can build an infinite number of functions $T \to [0, 1]$, then we can say that the set in the totality of its aspects and manifestations never becomes exhausted in any of its particular phenomenal aspects. The set is massive, fixed, complete in itself, totally and wholly given. It is devoid of potency and becomes, roughly equivalent to the inert world of objects and things. A fuzzy set is fluid and vacuous rather than fixed and full. It is characterized by incompleteness, potency, and lack. As 'set' is roughly equivalent to the inert and solidified world of objectivized reality, so fuzzy set generally corresponds to the human consciousness. These two modes of being, however, are not granted an equal ontological status. The set is both logically and ontologically prior to fuzzy set. The latter is dependent upon the former for its origin. The fuzzy set is inconceivable without the set and is derived from it through ordering; it thus constitutes an ordering of set. The fuzzy set is a borrowed being that emerges from the set by virtue of ordering. It is through man or human thinking that fuzziness comes into the world. The consequence of the primordial rupture of the fuzzy from the nonfuzzy is the introduction of abstraction. Concepts are dependent upon the objective reality in a way that reality is not dependent upon concepts. Ordering is not an abstract idea complementary to being, nor can it be conceived outside of being; it must be given at the heart of being. Ordering demands a host, possessing the plenitude and full positivity of being from which it borrows its power of representation.

We have observed that it is through man that fuzziness comes into the world. The question then arises: what is it about the being of man that occasions fuzziness? The answer is synthesis, freedom to go upward to higher synthesis looking for structural stability.

In many situations there are several choices but not all are equally acceptable. The degrees of acceptability depend on the precise context in ways that cannot be or would be very inconvenient to have expressed in advance. The reason that vague descriptions are used is then fairly clear: to achieve robustness and compactness.

3.3. DECISIONS

Divide and conquer seems to be the first principle of scientific attack. Breaking a complex system into tractable pieces is often the only way to understand it. Complex systems are understood by first explaining the parts and then understanding the way they fit together. In addition to an ability to encode specific facts in simple concepts, a semantic system needs some facility for grouping sets of similar facts into units, allowing these facts to be represented collectively through some sharing mechanism, and to be conceptualized as an integrated whole.

A convenient method for organizing a semantic system is to group together like objects, allowing similar objects to be thought about and talked about collectively. The scheme is hierarchical, in that some concepts may be subconcepts of more general concepts. The lower a concept is in the concept hierarchy, the more alike its members must be. The conceptual system serves the important purpose of spotlighting similarities among objects and compressing redundant information by recording the common characterizing property.

Let us follow the reasoning of Mesarovic[71]. He says that the simplest way to describe subsystems is in terms of functions $f_i : X \to R$, where X is the set of objects and R is the set of real numbers. A general decision problem can be formulated as follows: given a subset $S \subset X$, find $\hat{x} \in S$ such that for all $x \in S$

$$f_i(\hat{x}) \leqslant f(x)$$

The set X is the decision set, while the set S is the set of feasible decisions. An element $\hat{x} \in S$ satisfying the above inequality for all $x \in S$ is a solution of the optimization problem specified by the pair (f_i, S). When the function f_i is specified by two functions $M : X \to Y$ and $F_i : X \times Y \to R$, then

$$f_i(x) = F_i(x, M_i(x))$$

and M_i is referred to as the model of a control process and the function F_i is referred to as a performance function.

Since the subsystems operate in order to achieve their own individual goals, a conflict generally develops among them. Absence of conflict means that a decrease in each of the partial functions does not cause an increase in the overall function g. Therefore $g(x) = \min g(x)$, whenever the object $x \in X$ is such that $f_i(x) = \min f_i$, $i = 1, \ldots, n$. In other words, the equation

$$g(x) = G(f_1(x), \ldots, f_n(x))$$

holds for each object x in X when G is monotone, i.e. order-preserving.

Let us see now what is happening if g and f_i are fuzzy sets, that is $g, f_i : R \to [0, 1]$. If we denote by $\wedge$ the operation of taking the minimum, then we can show that if

$$g(x) = f_1(x) \wedge f_2(x) \wedge \ldots \wedge f_n(x)$$

and $\hat{x} \in R$, then $g(\hat{x}) = \inf g(x)$ when $f_i(\hat{x}) = \inf f_i(x)$. The proof is quite simple. Let $a = f_1(\hat{x}) \wedge \ldots \wedge f_n(\hat{x})$ and $b \inf g(x)$. We must show that $a = b$. Clearly $b \leqslant a$. Conversely, for all i, $f_i(\hat{x}) \leqslant f_i(x)$. Therefore, for all $x \in R$

$$a = f_1(\hat{x}) \quad \ldots \quad f_n(\hat{x}) \leqslant f_1(x) \quad \ldots \quad f_n(x) = g(x)$$

and $a \leqslant \inf g(x) = b$.

So far we have had a static image of what is called a decision. The decision was viewed as a final judgement, a settlement or, more precisely, an attainment.

We shall consider now decision-making as an act. The term 'act' covers both processes and attainments. It is the mark of a process that it has duration, but an attainment, although datable, is not time-consuming. Decision-making could be defined as a struggle to resolve a dilemma of conflicting objectives (Zeleny[72]). This distinction between compaigning and winning is signified by the terms 'process' and 'attainment'. The set of decision criteria is evolving and changing during the process of decision-making. We speak about unfolding a decision; partial decision-making includes a directional adjustment of the decision situation. It consists of adding criteria. The new criteria are combined in order to have, finally, only one criterion. Problem-solving is thinking. The activity is analogous to building a bridge from the mainland to an island; the mainland is the body of understanding a person already possesses, and the island is some factual knowledge that it is as yet unassimilated by the mind. The aim of thought is to bring the new into intelligible relations with the old. Unfolding a decision is a criteria-melting process. First, there is a sense of conflict; this provides the decision-motivating tension, a period of dissatisfaction with the *status quo* of a current situation. The underlying source of predecision conflict is structural instability. I think that the source of predecision conflict, which triggers the decision process, is the infeasibility of structural stability. Because of the experienced conflict of different criteria, one starts searching for a new representation that is structurally stable.

Let us consider a criterion $f_1 : X \to [0, 1]$. We say that this criterion is structurally stable if, roughly speaking, the shape or the ordering is not affected by small changes in the set X, i.e. if a slight change in the carrier X yields only a slight change in the membership function f_1. Now, if a new criterion $f_2 : X \to [0, 1]$ is taken into account, i.e. the same carrier is viewed from a second point of view, one seeks an integration of the two descriptions f_1 and f_2; a new criterion $g = G(f_1, f_2)$ is constructed. It is the characteristic of a whole to bring diverse elements together into a higher unity. The notion 'man' is arrived at after considering such characteristics as height, colour, musical ability, and the like; initial properties are reconciled in a higher synthesis. One has a collection of descriptions and, according to simple interaction rules between them, a new description emerges spontaneously as a consequence of those interaction rules. A dynamic series of descriptions is formed which continually affects each other. As a result of this movement a compact description is achieved.

When we try to understand a reality we customarily pass from the concept to the thing, rather than the reverse. Now, criteria are abstractions; they portray what is common to objects. What we are doing is to try to capture an object by putting criteria together. A criterion can only circumscribe around an object a frame that is too large and does not fit exactly. Realizing this in the case of any one criterion, we add another, which is also too large but which partially overlaps the previous frame and thus cuts down the area within which the object is to be found; we are moving around the object X. Because of spatial perspective an object appears different from various points of observation. To identify the object with any one of these appearances would be a mistake. All such knowledge is relative and partial; there could be an infinity of such criteria. Perhaps the most noteworthy feature of the decision unfolding process based on synthesis is the stability of this movement.

It is quite natural to think of this change as a succession of states in which each state holds the stage for a while before it is followed by a successor. This change is a process in which states are fuzzy sets. With the notation $f_1 = x_k, f_2 = u_k$ and $g = x_{k+1}$ we can write the standard state equation

$$x_{k+1} = x_k \wedge u_k$$

Clearly, in this framework we can handle control problems. Since small changes in control inputs u_k have negligible effects upon the final state, we say that the fuzzy system is stable. But this is another story, the topic of section 4.3.

It is proposed here that the principle of homeostasis, the law that organisms seek to maintain inner equilibrium, can be expanded into a major generalization in the field of decision theory. This principle seems to have value in illuminating decision-making considered both as a universal and as a unique phenomenon. The concept of 'criterial stability' must be introduced now if we are to get a clear picture of the role played by *homeostasis* (*homoios*, of the same kind + *stasis*, condition) in decision-making. In attempting to reduce the variability of external stimuli the decision-maker must deal with structurally stable logical objects.

Criteria and objectives are concepts; they eliminate the need for treating each situation as new. In ordinary language they are called habits. The use of criteria as coordination devices is limited, however; it is limited to those situations that can be anticipated. As the decision-maker faces new and different situations, the use of individual criteria must be supplemented by other integrating devices. The innovative or learning step is necessary for the decision-maker to remain stable under unusual, or new, or nonroutine environmental conditions. Control, then, includes not only regulation of routine events but also innovation. Therefore, the concept of structural stability will aid in distinguishing between systems that achieve stability under specific constant conditions and those that can learn or evolve new structures and behaviour in order to remain stable under changing conditions; the latter is the focus here. Another way of expressing the above is to say that the open systems face the fluctuating environment through processes of learning and innovation. The emergence of stable, enduring patterns of decision-making is in part a process of conflict resolution. It is the supreme tendency of the decision-maker to get rid of tensions and drives and come to rest in a state of equilibrium. Hence it can appear that the unfolding of a decision is an attempt to seek this point or, as it is sometimes expressed, the system is seeking the goal. Every criterion is a model in the broad sense of the word; that is, a conceptual image intended to reflect certain aspects of reality. One of the important aspects of the modern changes in scientific thought is that there is no unique and all-embracing criterion. The fact that similar peak situations may develop from different initial starting points is a striking example of equifinality. To summarize, repeated disequilibration results in continuous modification of the global representation. Dynamic homeostasis involves the maintenance of inner stability by establishing a unique criterion, the new equilibrium.

3.4. CONFLICT AND SYNTHESIS

By learning, I mean, the change in the state of an organism as a result of receiving information. With this definition it is obvious that science is equivalent to learning.

Due to the complexity of reality, the scientific approach of a problem has always followed the rule *divide et impera*. For a long time it was assumed that assembling the parts would allow one to understand the whole. The systemic approach reacted and warned that the interaction of parts might determine behaviour. A thorough study of psychological processes has detected hierarchical organizations. It was found that assembling does not mean juxtaposition, but aggregation. The study of hierarchical systems allowed the understanding of linguistic processes closely tied to the process of knowledge. With an hierarchical representation one saves description; natural language allows compact description, using a hierarchy of concepts.

Not long ago a concept was modelled as a set of objects having common properties. Recently, an important interest was shown by linguists, psychologists and philosophers for considering imprecision in quantifying the meaning. Recent researches in semantic theory raised the question of whether the meaning of a concept was always precise. It was established that, in most cases, the meaning is fuzzy; that is the borders of applicability of the concept are not clear: between applicable and non-applicable there is a degree of applicability.

The study of conceptualization, that is the process of concept aggregation, was stimulated by modelling the concept as a fuzzy set, i.e. a function in a universe of discourse with values in a lattice.

A concept is no more viewed as a set of objects described by an attribute; such a view proved to be unrealistic. It was found more adequate to represent it as function; all the objects of the universe have, to a smaller or a greater extent, a certain property. In other words, the objects of the universe can be ordered following various points of view.

Taking into account several points of view, it is easy to emphasize their mutual conflict. We are not speaking of contradiction, because we do not deal with dichotomies; we consider only differences in ordering. If we want to take into account simultaneously all the points of view, we aggregate these points of view. From the concept 'bird' and the concept 'mammalian' we build the concept 'animal'. In other words, we overlapped the common; we retained the essential so as to use one single label.

The hierarchical system arises by devising abstracts. From concepts, decisions and branches of knowledge there arise general concepts, global decisions and marginal branches. The building of the general concepts denotes a rising above hierarchy. Aggregation of primitive concepts, of individual decisions and of branches is accompanied by synthesis, coordination and generalization. In moving from one level to another, a unitary view is essential. The move, the jump from one level to another, is dictated by conflict. Its solving means compatibility and harmonization, and leads to descriptive compaction.

Between the fuzzy sets characterizing concepts at one level, and the fuzzy sets that synthesize them at the level above, there is a link that must be monotone.

An elevated representation of the process of concept aggregation, of fuzzy set composition, can be made in the frame of the theory of categories. Considering the category having as objects fuzzy sets, and the morphisms which are their transformations—defined as functions on carriers—the jump from a level of devising abstracts to another means the same as the building of a projective limit or, in the case of a tree, a pullback. Hence learning may be equivalent to the building of a series of projective limits.

A learning process develops in time and is therefore a form of movement. It is natural to ask ourselves what the state of the system is. At first sight it may seem to be a matter of quantity of information. This would mean that, as time elapses, the entropy of the learning system decreases; at a closer look, it is the modelling of the learning system as a fuzzy system, with the state equation given by the equality: a fuzzy set on a level is a function of the fuzzy sets situated on the lower levels, which shows that as the standard (astronomical) time increases, the aggregation level becomes higher, and, therefore, the degree of imprecision increases. A philosophical theory is more general than a narrow branch which supplies the former with arguments.

The question of the practical utility of the proposed learning model arises immediately. For me it meant a useful methodological frame for directing our researches relating to management theory.

The harmonizing of goals with resources by placing decisions on higher levels involves a global approach. Here is an example—In most practical optimization cases, using linear programming models, one does not obtain a solution; the domain of admissibility given by the constraints is usually empty. In other words, the burdensome operation of using an expensive computer, after having made an equally expensive analysis, might lead to no practical conclusion. And this is because the planning operation reduces itself to a linear programming exercise, and the excess of care leads to an exaggeration of the constraints. But planning means something different: a rational way of adjusting constraints in order to get a solution. This can be made by reformulating the classical model, relaxing the constraints (i.e. converting them into cost functions), and reformulating the problems at a higher level by taking into account the confluence of the new functions resulting from greater flexibility.

These observations are in accordance with the present tendencies of researchers in the field of management science. Until now, goals and means were always emphasized; recently, this division was revised. It was found that, in fact, neither goals nor means can be determined independently of each other. The new goals arise on the basis of available alternatives which, in turn, are adjusted and generated according to existing goals. The modelling of decision-making and economic behaviour is now approached by taking into account the fact that the dichotomy 'means-goals' is not proper to real decision situations. Decision-making is characterized by a process that eliminates conflict. We speak now about decision dynamics, which is in fact a learning process. Thus, the theory of a displaced ideal is reduced to a theory of an estimation movement, which can be perfectly modelled in terms of fuzzy systems.

The global approach is interesting, and the proposed learning model can lead to practical conclusions also for developing information systems.

The organization of goal hierarchies, or simply of optimization problems, permits a new and profitable approach of the principle of the internal model, recently emphasized by studies in cybernetics. This principle is formulated as follows: to manage a system, the decision element must include a model of the exosystem, i.e. of the environment, in which the managed system is situated. By environment one understands both the space of controls and the space of perturbations.

Now, because complex systems are represented by parts assembled by aggregation, along a tree-chart, every node of the tree must contain a model of the higher level. The condition is necessary and sufficient to secure harmony; ranking eliminates errors due to collateral information. For coordinated management we do not have to know all the possible interrelations of the basic elements if the managers have a general view; important information savings are thus achieved. It is true that all is linked to all—an old slogan of the systemic approach—but it is by means of the initial object; hence we are dealing with a compact representation of interrelations. The practical importance of this remark is not at all negligible.

The hierarchical organization is well known. Galbraith[73] for instance, says that the organization design problem is to create mechanisms by which an integrated pattern of behaviour can be obtained across all the interdependent groups. The simplest method of coordinating interdependent subtasks is to specify necessary behaviour in the form of rules in advance of their execution. The primary virtue of rules is that they eliminate the need for further communication among the subunits; but the use of rules as coordination devices is limited to those job-related situations that can be anticipated. As the organization faces new and different situations, the use of rules must be supplemented by other integrating devices. The violation of known situations usually requires additional decision-making and, hence, additional information-processing; the additional decision-making takes place through the hierarchy. Problems are handled on an 'exception base' and they are raised to higher levels of the hierarchy for resolution. In this manner the behaviour of the interdependent subunits remains integrated. However, as the organization performs more uncertain tasks, the hierarchical channels become overloaded. The volume of information processing can overwhelm an organization behaving in a mechanical fashion. The organization must adopt a strategy to reduce the information necessary to coordinate its activities in order to reduce the number of exceptions referred upward through hierarchical channels. An organization can reduce the number of exceptions that occur simply by reducing the required precision. Over periods of time the department is independent of other departments and can act autonomously. This means that the decision-maker does not have to collect information about first and second effects of his decision. Fuzziness absorbs them; uncertainty absorption takes place when inferences are drawn from a body of evidence, and the inferences, instead of the evidence itself, are then communicated.

Goguen[74] noticed that there may be several explanations for the fact that large complex systems, whether natural or man-made, frequently exhibit hierarchical organization. The one he finds deepest and most satisfying is that it facilitates economical specification. Even when a system is not hierarchically organized we may employ such organization to simplify our description of it. In fact, hierarchical structure is natural enough to man so that many presumed hierarchies are more likely to reflect properties of the description and describer than those of the described.

It is argued by Pollatschek that hierarchical systems are necessarily fuzzy and lend themselves, in principle, to analysis by the techniques of this theory, properly adapted for this purpose.[75]

Cybernetics deals with purposeful systems controlled by error, i.e. by the difference between the state of behaviour of a unit at any time and the final state interpreted as purpose or ideal state. Each next state is a goal, and goal-seeking means dynamics, a move from one state to another. The motivation of this movement is the difference between the desired and the actual. In this book, purposefulness means the will to eliminate ambiguity introduced by partial description of reality. Dynamics aims at conflict resolution and is developing in the hierarchy of concepts. Caught up in ambiguity, there is only one possible outcome: to resolve the ambiguity.

Chapter 4

FUZZY SETS

There has recently developed quite an extensive literature on the theory of fuzzy sets. This chapter is an attempt to summarize some aspects of that literature, emphasizing the new trends. A fuzzy set, in this book, is a function $X \to [0, 1]$ from a set X, called the carrier, to the unit interval of real numbers. A later development considers linguistic truth values rather than precise numerical values in the unit interval. Thus, in the linguistic approach a fuzzy set is a function $X \to V$, where V is an ordered set of linguistic values[76]. Details about this later development can be found in the impressive works of Zadeh[77].

A key point in the study of fuzzy sets is their organization as a lattice. The idea of movement in this lattice is based on the dynamics induced by the intersection of fuzzy sets; this is a stable movement. Any lattice can be viewed as a particular category that is a formal system, and for this reason we shall speak further about systems in a category.

A *topos* is a category satisfying certain conditions. The notion of topos could be viewed as a conceptual form of higher order language, as a world in which higher order notions could be interpreted. A topos has the features of a syntactic object, which, though of a finite nature, is able to speak about infinite ideas.

The aim of this chapter is to give an intuitive sketch of the fuzzy set theory. The mathematical difficulties that arise in converting the sketched programme into rigorous proofs are isolated as a number of relatively plausible assertions that can be taken on trust without interrupting the flow of the argument. However, it is hoped that a little insight into the algebraic theory of fuzzy sets may prove valuable for those who are unhappy about taking the results entirely on trust.

4.1. FUZZY SETS AS FUNCTIONS

Zadeh[6] defines fuzzy sets as being functions from some ordinary set X to the unit interval [0, 1]. Given sets X and [0, 1], we shall use the notation

$$f: X \to [0, 1]$$

to denote the function defined for all x in X and with values in [0, 1]. The assumption underlying fuzzy sets is that the transition from membership to nonmembership is seldom a step function; rather there is a gradual but specific change.

In nonfuzzy set theory, a characteristic function specifies which elements are members of the set, i.e. for which elements the statement $x \in X$ has a truth value of 1. In fuzzy set theory the grade of membership and the corresponding truth value of the proposition $x \in X$ may take on any value in the closed real interval from 0 to 1.

A fuzzy set is empty if it is the constant function zero, and two fuzzy sets are equal if they are equal as functions. The fuzzy set f is a subset of the fuzzy set g if, and only if, for all x in X $f(x) \leqslant g(x)$

The union of the fuzzy sets f and g is the function $f \cup g$ defined by $(f \cup g)(x) = \max\,[f(x), g(x)]$, i.e. the smallest fuzzy set having both f and g as subsets. The intersection of the fuzzy sets f and g is the function $f \cap g$ given by $(f \cap g)(x) = \min\,[f(x), g(x)]$, i.e. the largest fuzzy set that is a subset of both f and g. The complement of a fuzzy set f is the function defined by $f'(x) = 1 - f(x)$.

Using these definitions, both De Morgan's laws and the usual distribution laws hold

$$\begin{aligned}(f \cup g)' &= f' \cap g' \\ (f \cap g)' &= f' \cup g' \\ f \cap (g \cup h) &= (f \cap g) \cup (f \cap h) \\ f \cup (g \cap h) &= (f \cup g) \cap (f \cup h)\end{aligned}$$

However, although the fuzzy sets constitute a distributive lattice with 0 (the empty set) and 1 (the constant function 1), the operation of complementation does not make this lattice into a complemented lattice. While it is true that such pleasant properties as De Morgan's laws and the law $f'' = f$ hold, it is certainly not always the case that $f \cup f' = 1$. For example, if $f(x) = a$, and a is neither 0 nor 1, then, as can be verified from the definitions

$$\begin{aligned}(f \cup f')(x) &= \max\,(a, 1 - a) \neq 1 \\ (f \cap f')(x) &= \min\,(a, 1 - a) \neq 0\end{aligned}$$

A fuzzy relation is a fuzzy subset of $X \times X$. The composition of two fuzzy relations $f(x, y)$ and $g(x, y)$ is the fuzzy relation

$$h(x, y) = \sup \min [f(x, v), g(v, y)]$$

Thus, for any particular pairs (x, v) in f and (v, y) in g indicating that (x, y) should belong to h to some degree, we assign (x, y) the minimum of the corresponding degrees of $(x. v)$ and (v, y), and then assign as a final value to (x, y) supremum of these minima.

Chapin[78] developed a Zermelo-Fraenkel-like theory based on a membership relation, eliminating the need to have a predetermined theory of ordinary sets on top of which the fuzzy sets are built as a superstructure, by starting out axiomatically *ab initio*, as it were, assuming only elementary logic.

Membership in a fuzzy set is specified by a mapping, performed by enumeration, by a function, by an algorithm, and the like. Once this mapping is specified, the fuzzy set can be used in inferences and can be modified by operations (Zadeh[79]). For instance, if the concept 'small' is represented by the function f and the concept 'large' by the function g then, 'not small' is represented by $1 - f$ and 'not large' by $1 - g$, 'very small' by f^2, and 'very large' by g^2, 'not very small' by $1 - f^2$ and 'not very large' by $1 - g^2$, 'very, very small' by f^4, and so on. In addition to the operations of union, intersection, complement algebraic product and algebraic sum, new operations called 'bounded sum' and 'bounded difference' were defined by Zadeh[9] to investigate fuzzy reasoning. Mizumoto[80] investigates the algebraic properties of fuzzy sets under these new operations.

DeLuca and Termini[81, 82] have introduced in the lattice of fuzzy sets two functionals P and d called respectively 'power' and (logarithmic) 'entropy'. The power defined for all fuzzy sets f in the lattice as $P(f) = \sum_x f(x)$ generalizes the cardinality of a classical set. The entropy, which vanishes in the classical case, is a measure of the total amount of vagueness of a fuzzy set and is defined for all f in the lattice as $df(f) = \sum_x S(f(x))$ where S denotes the Shannon function $S(x) = x \ln 1/x + (1 - x) \ln [1/(1 - x)]$.

Nguyen[83] examines the concept of linguistic probability introduced by Zadeh[77] and explores its relations with fuzziness in measure theory.

Kandel[84] has published results concerned with fuzzy logic. He has shown that every fuzzy form can be expressed in disjunctive or conjunctive normal form using the operations max and min. Digital realizations of the operations max and min and the realization of fuzzy forms is considered by Srini[85]. A new approach, based on many valued-logic of the theory of fuzzy sets was published by Moisil[86] in 1975. Fuzzy algebras were studied recently by Ponasse[87] and Klement[88] universes of fuzzy sets are studied by Gottwald[89].

In the following I shall give a geometrical intuitive sketch of the way in which fuzzy sets are to be understood. This sketch was suggested by Flondor.[90]

For any set X we denote by $F(X)$ the set of all fuzzy sets having X as carrier, and by $P(X)$ the power set of X, i.e. the set of all subsets of X, including the empty set. Each element a of X will be represented as a subset consisting of a single element. Instead of these singletons we can use their characteristic function $\psi: \{a\} \to [0, 1]$ with $\psi(a) = 1$ and $\psi(\emptyset) = 0$. With this notation one can represent $P(X)$, when X contains two elements a and b, as four sets (0, 0), (0, 1), (1, 1), (1, 0). Every such set can be represented as a point in a bidimensional space as below.

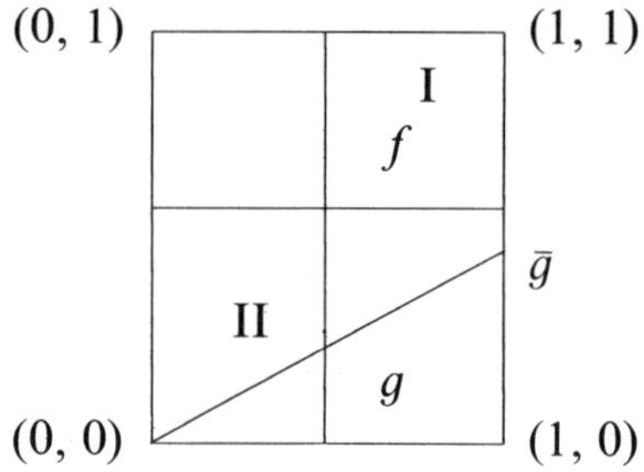

We shall now pass in review a number of basic notions. For instance, we have already defined a fuzzy set f as being a subset of a fuzzy set g if, and only if, $f(x) \leqslant g(x)$. The relation $\leqslant$ represents an inclusion. Area I represents those fuzzy sets containing f, and area II represents those fuzzy sets included in f. It is easy to see that (1, 1) includes all the points and (0, 0) is included in all the points.

To make this reasoning we have identified $F(X)$ by the Cartesian product $[0, 1] \times [0, 1]$. These conventions may sound unorthodox but are actually intuitive and permit an easy understanding of the attainability of $F(X)$. In the above pictorial representation $P(X)$ is indicated by four points and $F(X)$ by a surface. Let us go further. The height of a fuzzy set is the maximal value of the function. A fuzzy set is said to be normal if its height is unity. Clearly, in our representation, the boundaries of the square are normal fuzzy sets. Then, if we want to normalize a nonempty fuzzy set $\bar{g}$, a line from (0, 0) passing through g will cross the boundary in $\bar{g}$; the assignment $g - \bar{g}$ is the normalization. All the points along this line have the same normal fuzzy set.

The complement of a fuzzy set f is the fuzzy set $f' = 1 - f$. In our representation the complement of a point is its symmetry against the point $(\frac{1}{2}, \frac{1}{2})$ called, sometimes, the crossover point. This point is a fixed one; the complementation leaves it unchanged.

We now turn our attention to operations on fuzzy subsets. The first operation to be considered is the union. We may appreciate the nature of this operation by again considering our square.

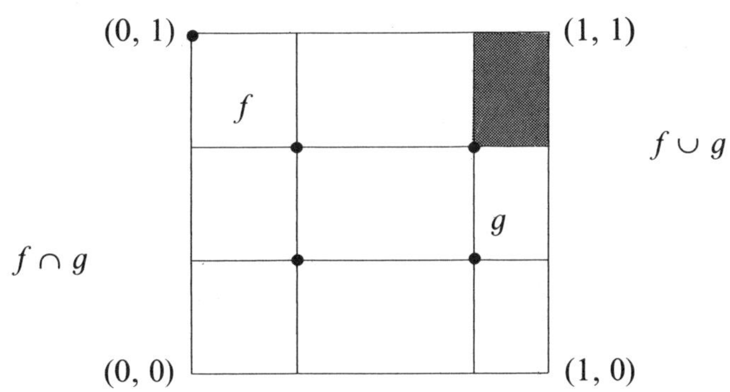

The union of the fuzzy sets f and g is the function $f \cup g$ defined as $f \cup g = \max(f, g)$. We are looking for the smallest fuzzy set having both f and g as subsets. If we recall the representation of inclusion, then, $f \cup g$ is the corner of the double-shaded region. Let us consider a family of fuzzy sets represented by a circle centred on $(\frac{1}{2}, \frac{1}{2})$. It is quite clear that $f \cup g$ is the union of this family and $f \cap g$ its intersection.

Note that $F(X)$ is a logical ring, that is, a set with two laws of composition and written as union and intersection. An ideal I in this ring $F(X)$ is a subset of $F(X)$ such that

$$f, g \in I \rightarrow f \cup g \in I$$
$$f \in I, g \subset f \rightarrow g \in I$$

In our square, any area determined at left by a point is an ideal. In the same manner we can construct a filter in the ring $F(X)$ as a subset of $F(X)$ such that

$$f, g \in F \rightarrow f \cap g \in F$$
$$f \in F, f \subset f \rightarrow g \in F$$

A fuzzy topology defined as a family T of fuzzy sets on X which satisfies the following conditions

$$\phi, X \in T$$
$$\text{if} \quad f, g \in T, \quad \text{then} \quad f \cap g \in T$$
$$\text{if} \quad f_i \in T \quad \text{for each} \quad i \in I \quad \text{then} \quad \cup f_i \in T$$

is a set of points in the representation square. For instance, a principal filter with ϕ or a principal ideal with X are fuzzy topologies. (There are also other points of view [91–94].)

We can speak also about trajectories in the square. We call a mapping $t: M \rightarrow F(X)$ a trajectory or a motion in the plane $F(X)$. If M is an interval in the real line the trajectory is a curve and may be regarded as having been traced out by the tip of $t(m)$ as m increases. A trajectory is said to be continuous if the function $M \rightarrow F(X)$ is continuous, $F(X)$ having the Euclidean topology ($F(X)$ is compact). If X has only two elements, the image of the continuous trajectory is a continuous curve in R^3. It should be stressed that the natural topology on $F(X)$, induced by R^2, allows one to work with a strong topology.

Within this framework we can speak also about fineness, observation, and observation operators (Chang and Zadeh)[95]. The degree of exactitude in our knowledge can be represented by the concept of fineness; if our knowledge improves, the fuzzy set becomes finer. If our knowledge becomes exact, the fuzzy set becomes a corner of the square. Since our knowledge is improved by observation, the effect of an observation operator on a fuzzy set is to make it finer. In this context an observation on a fuzzy set means its renormalization.

4.2. THE LATTICE OF FUZZY SETS

Let us consider again the set $F(X)$ of all fuzzy sets. Let f, g be two fuzzy sets, f being included in g. $F(X)$ together with the relation of order $f \leqslant g$ is an example of a partially ordered system consisting of a nonempty set $F(X)$ and a binary relation $\leqslant$ in $F(X)$ such that the following conditions are satisfied for all $f, g, h \in F(X)$

$$f \leqslant f$$
$$\text{if} \quad f \leqslant g \quad \text{and} \quad g \leqslant f \quad \text{then} \quad f = g$$
$$\text{if} \quad f \leqslant g \quad \text{and} \quad g \leqslant h \quad \text{then} \quad f \leqslant h.$$

The relation $\leqslant$ is a partial order in the set $F(X)$, and $F(X)$ is said to be partially ordered by the relation $\leqslant$. Although the set $F(X)$ is only the domain of the partial order $\leqslant$, we shall follow the usual custom of identifying $F(X)$ with the partially ordered set. This partially ordered set may be diagrammatically represented, using the concept of covering. If f and g are elements of $F(X)$, $f < g$, and there is no element $h \in F(X)$ such that $f < h < g$, we say that f is covered by g, or g covers f, and we write $f < g$ or $g > f$. Now, draw each element in $F(X)$ in such a way that the element g lies above the element f whenever $f > g$, and connect them by a line segment. We call the resulting figure a diagram and say that the partially ordered set $F(X)$ is represented by this diagram as below.

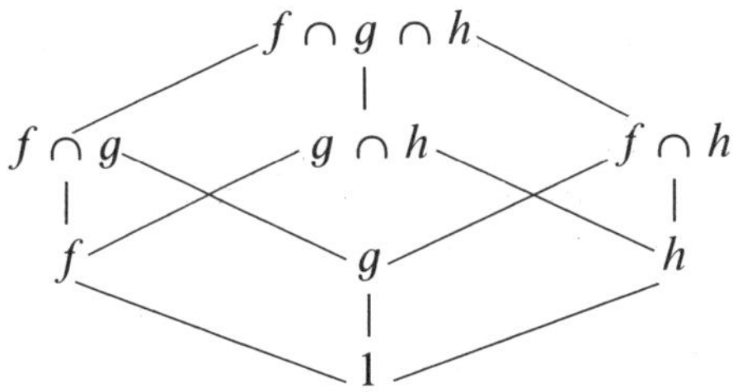

where 1 is the function having all the values equal to one.

If S is a subset of the partially ordered set $F(X)$, there is at most one element $a \in S$ such that $a \geqslant f$ for each $f \in S$; such an element is called the greatest element of S. Similarly, if S contains an element b such that $b \leqslant f$ for all $f \in S$, then b is called the least element of S. An element a in $F(X)$ is said to be an upper bound of a subset $S \subset F(X)$ if $a \geqslant s$ for every $s \in S$. Similarly, a is a lower bound of S if $a \leqslant s$. We say that an element $a \in F(X)$ is a least upper bound of S if a is an upper bound of S and $a \leqslant s$ for every upper bound s of S. Similarly, b is a greatest lower bound of S if b is a lower bound of S and $b \geqslant s'$ for every lower bound s' of S; *lub* and *glb* are unique.

We have said that the inclusion of fuzzy sets means ordering. In other words, we say that there is a unique *morphism* from f to g and we write

$$f \to g$$

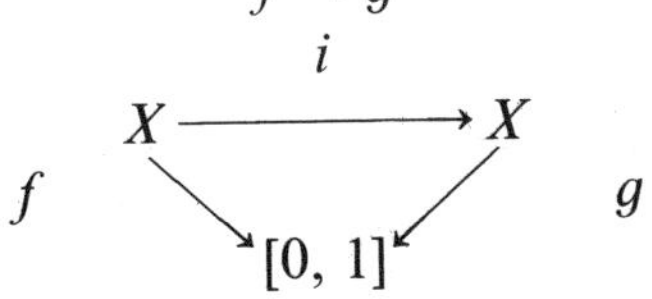

where i is the injection. In order to have $i(x) = x$ we must have satisfied the condition $g \circ i \geqslant f$. Using this language we say that a is a lower bound for $S \subset F(X)$ if, and only if, there exists a morphism $a \to s$ for each s in S. Now if we define a terminal, the object t, such that for every object s there is a unique morphism $s \to t$, then glb is a terminal.

Generalizing this case we can imagine a morphism m instead of the injection i such that $g \circ m \geqslant f$. A step further is to consider the situation of having a family m_i instead of a unique morphism m. We shall consider now a diagram that is a directed graph whose edges i, j are labelled by the morphisms m_i. A cone for a diagram is a family of morphisms $f \to g_i$ from a single fuzzy set f such that $f \to g_i \to g_j = f \to g_j$ for every $g_i \to g_j$. A morphism from a cone $(f \to g_i)$ to a cone $(f' \to g_i)$ is a morphism $f \to f'$ such that $f \to f' \to g_i = f \to g_i$. If for all cones $(f' \to g_i)$ on the diagram there is a unique morphism $f' \to f$ such that $f' \to f \to g_i = f' \to g_i$, then the terminal cone is the limit of the diagram.

To summarize, we are dealing now with a collection of logical objects, the fuzzy sets, $|F|$, and for two such objects $f, g \in |F|$ a set denoted $F(f, g)$ which is the set of morphisms of f into g, and for three objects $f, g, h, \in |F|$ a law of composition

$$F(f, g) \times F(g, h) \to F(f, h)$$

satisfying the conditions

> associativity: given morphisms $f \xrightarrow{A} g \xrightarrow{B} h \xrightarrow{C} j$ we have $(AB)C = A(BC)$ identity: for each object f, there exists a morphism $1_f : f \to f$ which satisfies $1_f A = A$ for every morphism $A : f \to g$ and $B 1_f = B$ for every morphism $B : h \to j$

But this is precisely the definition of a category. Although this book will not use category theory as a systematic mathematical tool, many of the notions that we shall consider lead to categories and it will be convenient and useful to identify them as such and use an appropriate terminology. For more on category theory, *see* MacLane[96], or Arbib and Manes[97,98].

A categorical theory of fuzzy sets was presented for the first time by Goguen[99]. He noticed that categories are particularly appropriate because they allow us to discuss relations between and operations upon concepts without committing ourselves about their internal structure. Loosely speaking, a morphism $A : f \to g$ models part of the structure of f with part of the structure of g. This assumption could be validated by showing that the structure can be recovered from just the system of morphisms and compositions. This took the form of a representation theorem that any abstract category satisfying certain (purely category theoretic) axioms is equivalent to the category of all objects of the original structure, with appropriate structure-preserving morphisms. We then say that axioms categorize the structure, or are categorical. Goguen emphasizes that representation theorems may also have a broader philosophical significance as justification for the systems theoretic methodology of science now gaining prominence. Here one seeks to discover internal structure simply by observing external interrelations and interactions. This methodology is frequently contrasted with the more traditional approach which seeks structure by dissection or decomposition.

Now, let us consider two morphisms $A: f_1 \to g$ and $B: f_2 \to g$ we say that the commutative diagram

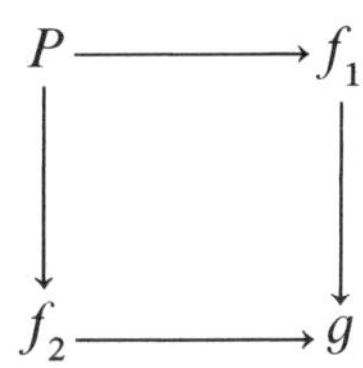

is a pullback of A and B if it has the property that any commutative diagram

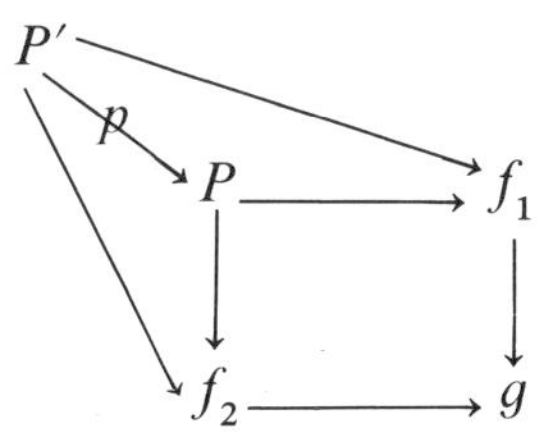

can be completed by a unique p.

Clearly, the pullback is the limit of the diagram

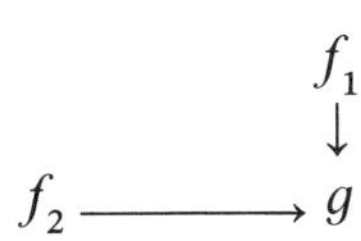

The relationship of fuzzy system theory to dialectics should be fairly obvious. The first dialectical law of the transformation of quantity into quality is very close to, though not exactly synonymous with, the transformation of many into one. Within the dialectical framework, a change of a quantitative nature, from g to f_1 and f_2, always leads to a qualitative transformation P. Such dialectical exposition is essentially identical to the fuzzy system theory presented in this book.

In dialectics the concept of 'development' is distinguished from that of 'growth' to indicate that real change is not a mechanical process. Development is said to take place through the mutual interpenetration of opposites. This developmental process is the internal content of the transformation of quantity into quality.

Now a few words about the categorical approach, which is an extension of that of the classical lattice. Before category theory came along, mathematicians expressed things by using the familiar classical language involving variables, constants, $=$, quantifiers such as $\exists$, $\forall$ and brackets like $\{(w_1, w_2, w_3, w_4) | w_1 + w_4 = w_2 + w_3\}$, and so on. Category theory arose because people, mostly algebraic topologists, found it convenient to use diagrams capable of expressing a great deal of information in a compact form. The reason for introducing commutative diagrams is that they allow us to represent many equalities between functions by one two-dimensional diagram which is easy to visualize, rather than spreading the information across many equations from which it is hard to retrieve. In fact, the whole strategy in category theory is to show that many results can be obtained in great generality if we chase arrows around diagrams rather than by using element-by-element computations. Diagrams can certainly clarify things, and for a long time it was the fashion among category theorists to express everything in terms of diagrams. However, proving this with diagrams, in fact working categorically, is a much more tedious business than the old classical language with elements and variables. What has happened recently is a swing back to the old-fashioned notation, but now reinterpreted so that it is valid not just for sets and functions but for objects and maps in an appropriate sort of category. For example, the bracket referred to above is a shorthand notation for the equalizer of two maps $W \times W \times W \times W \rightrightarrows W$ one described by $(w_1, w_2, w_3, w_4) \to w_1 + w_4$ by using $W \times W \to W$ and projection maps and so on in the obvious way, and other by $(w_1, w_2, w_3, w_4) \to w_2 + w_3$. Here w_i are not elements but abstract symbols. This sort of notation can vastly simplify working and obviates the need for huge diagrams.

4.3. STATE SPACE MODELS

If we consider the epistemological realm, then an evolutionary theory is required to show how something can become unlike its origin in kind. It is suggested in this section that fuzzy systems theory can be a useful model for the necessary dialectical movement between different descriptions of the ontological realm. A stable fuzzy system is supposed to describe a synthesis process. A fuzzy system specifies interaction rules between different descriptions. As a consequence of these rules structural stability can emerge spontaneously. Two descriptions are reconciled by pullback in a synthesis. Learning is viewed as the longing for totality, completeness, and wholeness; it seeks a unification of detailed existence in such a way that any particular would be a member of a totality. Synthesis means to go upward to higher freedom given by structural stability. Organic growth is viewed as a dynamic series of states that permeate and strengthen the descriptions which are its members. When small changes in partial descriptions have negligible effects upon the final description the fuzzy system is stable.

A common type of phenomenon observed in many areas of human experience is characterized by the production of a series of orderings on the same collection of objects. These orderings tend to be definitely fixed as regards their possible distinct forms but haphazard as to the order in which one form follows upon another. Despite the absence of any discernable law of serial order, there usually exists a striking similarity between the compositions of any pair of long records of orderings. The observed regularity is the empirical basis for the so-called laws of abstraction or transcendence which are a subject matter of the theory of fuzzy systems.

The essential mathematical properties fuzzy systems must possess suffice to establish the theory of abstraction, on an axiomatic basis, as a rigorous branch of pure mathematics. Since the axioms appear arbitrary when stated without reference to their rational background, we shall indicate something of their conceptual origin by sketching an approach to the subject.

We shall denote by $F(X) = \{f: X \rightarrow [0, 1]\}$ the set of all fuzzy sets defined on the carrier X. It is known that $F(X)$ is a lattice such that all one or two element subsets of $F(X)$ have a greatest lower bound and a least upper bound. The existence of a unique $g$lb (lub) for each pair of fuzzy sets implies the existence of a relation from $F(X) \times F(X)$ into $F(X)$. This relation is a function, since the glb's (lub's) are unique and the mapping is one to one.

In this framework, three fundamental elements that characterize a dynamic process can be discovered. The elements are stage, state and transition. It is quite natural to think of lattice $F(X)$ as describing a change, as a succession of states. Here the inputs are given by genesis, dynamics is given by conflict resolution or synthesis, and homeostasis means movement (kinesis) toward equilibrium (stasis). The state description consists of a transition function that relates the current state with the input to obtain the state next stage. Two systems are therefore defined by the operation which consists in obtaining the glb of two elements and denoted $\wedge$, and the operation which consists of obtaining the lub of two elements and denoted $\vee$.

We now use these ideas to motivate some general categorical definitions. Consider three fuzzy sets A, B, C, related by two monomorphisms $A \xrightarrow{x_k} B$, $C \xrightarrow{u_k} B$ and the projections $p_A : A \times C \rightarrow A$ $p_C : A \times C \rightarrow C$. We say that $e : P \rightarrow A \times C$ is an equalizer for $x_k p_A$ and $u_k p_C$ if $x_k p_A e = u_k p_C e$ and if $e' = P' \rightarrow A \times C$ is such that $x_k p_A e' = u_k p_C e'$ there is a unique $m : P' \rightarrow P$ making the diagram

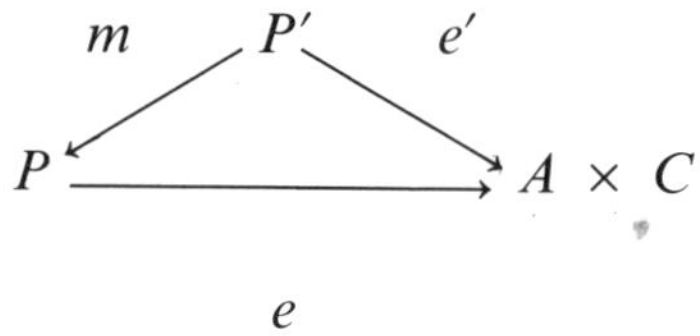

commutative. The equalizer defines the pullback

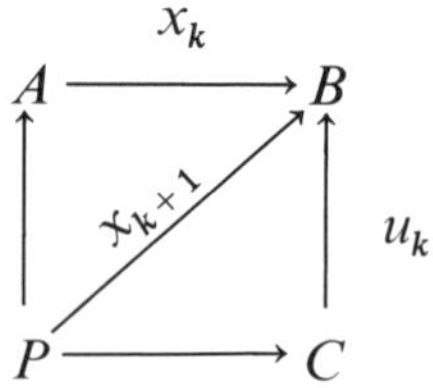

where x_{k+1} is a monomorphism $x_{k+1} = x_k \wedge u_k$.

We observe that x_{k+1} can be made arbitrarily small by an appropriate choice of x_k. The equilibrium of the system is thus stable. The equilibrium is not only stable but attractive since the system tends to ϕ, the empty set, as k increases without bound. Stability is a property of the system as a whole related to the coördination of the actions between the parts. But this is a synthesis that is never completed, is always on the way, an incressant striving that is never finished. Synthesis involves the maintenance of structural stability by reducing the disturbing effects of inputs. The pullback does not simply restore the prior equilibrium. A new, more complex and more comprehensive equilibrium is established. Here stable means remaining, or able to remain, unchanged in form under conditions tending to cause such change.

The assertion that a fuzzy system is stable can be interpreted by stating that conflict resolution means to go upward to a higher freedom than is given by structural stability, which means possibility. I do believe that within such a framework the reader could hope to deal with practical questions successfully. The state of a fuzzy system governed by the pullback rule tends toward a point ϕ which is an equilibrium position; this point is such that the trajectory of any point near ϕ goes to ϕ. We shall say that ϕ is an attractor of the system; in other words, that the same final state may be reached from different initial conditions and in different ways; this is what is called *equifinality*.

Every stable system has the property that, if displaced from a state of equilibrium and released, the subsequent movement is so matched to the initial displacement that the system is brought back to the state of equilibrium. The pairing of the line of return to the initial displacement can be regarded as intelligence and as being peculiar to humans.

We shall now devote a little time to the discrete nature of fuzzy systems. There are three fundamental notions in the theory of discrete processes: stage, state, and transition. There is always a scalar index, denoted by 'k', taking integer values, $k = 0, 1, 2, \ldots$ and we are concerned with the values of variables indexed by k, i.e. the notion of a discrete sequential process is connected with the assumption that the stage indexes have some kind of order. The state description of a discrete system consists of a transition function that relates the current state x_k to obtain the state of the next stage x_{k+1}. We shall use the notation

$$\mathrm{d}: S \to S \qquad s \mapsto \mathrm{d}(x)$$

for the transition function d of a discrete system to indicate that if the state at stage k is $s \in S$, then the state at stage $k + 1$ will be $\mathrm{d}(s) \in S$. We used the barred arrow to go from argument to value, in contrast with the straight arrow from domain to co-domain. Clearly, the transition function is a conditional statement which can be replaced by the concept of implication: if s_k, then s_{k+1}.

We can quite generally represent acts by functions mapping premises to consequences. Human (ethical) values are represented by an order of preference among the consequences, and human (cognitive) beliefs by an order of preference among premises. These two together determine norms of action as an order of preference defined for acts. This behaviour can be described in terms of fuzzy sets.

Just why dynamical systems are both interesting and challenging cannot be appreciated until one asks some control problems. If anyone is dissatisfied with the current state he may want to intervene and divert the system's path toward another state. It is then possible to see that in dealing with dynamical systems the fundamental problem is that at each stage it is necessary to make decisions and apply inputs to the system in such a way that the future system behaviour is satisfactory in some sense. The field of management has developed out of emphasis on change and on planned change. Therefore, we must consider the transition function

$$\mathrm{d}: S \times C \to S$$

where C is the input set. We use the notation $S \times C$ to denote the set $\{(s, c) | s \in S, c \in C\}$, i.e. the set of all ordered pairs (s, c) of which the first element comes from S and the second from C. Thus, we shall use the notation

$$(s, c) \mapsto \mathrm{d}(s, c)$$

to indicate that if the system is in some state from the set S at successive stages, and further, if the system is in state $s \in S$ and receives input $c \in C$ at the k stage, then the state at stage $k + 1$ will be $\mathrm{d}(s, c)$. This is precisely the case of fuzzy systems; for instance, the one defined by the operator.

$$: F(X) \times F(X) \to F(X)$$

We shall show, now, how a relationship induces a transition function. Consider a relationship $R: X \times Y \to [0, 1]$. To each R we can assign a function $w: X \to F(Y)$, where $w(x) = R(x, \cdot)$, and $F(Y)$ is the set of all fuzzy sets on the carrier X. The relation R induces an application

$$\mathrm{d}_R: F(X) \to F(Y)$$

If $f \in F(X)$ then $\mathrm{d}_R(f) = R(f) \in F(Y)$, where $R(f)$ is the image of f through R, i.e.

$$R(f)(y) = \sup_x (R(x, y) \wedge f(x))$$

Let us consider now the relation $R': X \times X \times X \to [0, 1]$ and the function

$$d_R : F(X \times X) \to F(X)$$

we shall also define a function

$$d' : F(X) \times F(X) \to F(X \times X)$$
$$d'(f, g) = f \times g \quad \text{the cartesian product}$$

Therefore we have

$$d_R^{d'} : F(X) \times F(X) \to F(X)$$
$$d_R^{d'} = d_R \circ d'$$

A fuzzy relation R' defines a relational system. Every finite relational system is a fuzzy automaton, a model of great linguistic interest if we view a language as an output of a system and the model consistent with special grammatical rules governing the construction of every possible sentence. Therefore, a fuzzy grammar is a fuzzy relational system.

Consider now the system

$$x_{k+1} = x_k + u_k$$

where x and u are fuzzy sets and the sum of two fuzzy sets is defined by

$$(x + u)(z) = \sup_{a+b=z} x(a) \wedge u(b)$$

i.e.

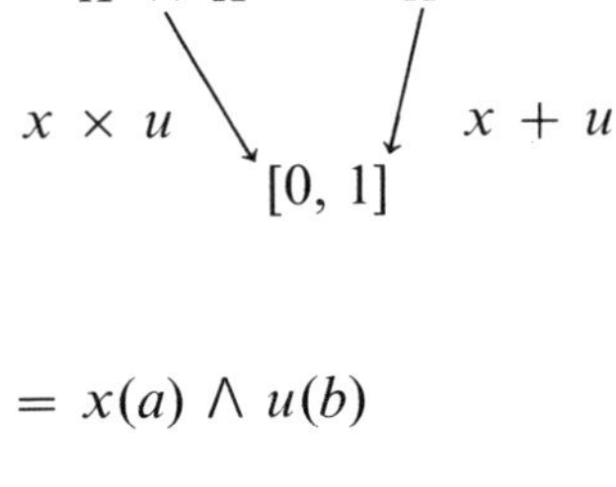

Since $(x \times u)(a, b) = x(a) \wedge u(b)$
we have the diagram

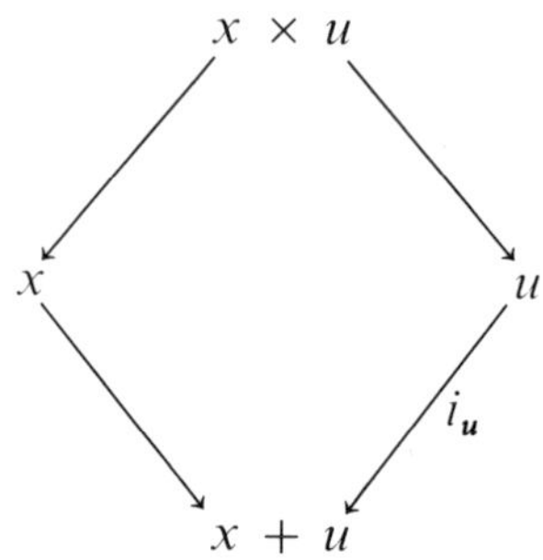

where i_u is the canonical injection

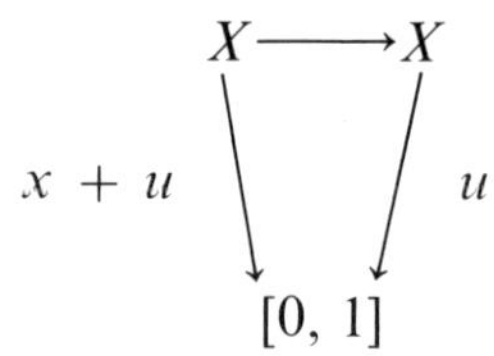

Having a structure on the carrier X we can define another structure on $F(X)$ beyond the lattice structure[100].

Planning means to estimate the desired path in an economic social or any other space, hence to guide. To devise reliable paths we should know the circumstances under which the imagined action will develop, including the actions of those who are not subject to our instructions; this is impossible practically. It is, hence, obvious that planning cannot be foreseen in detail. In this case a probabilistic approach is not adequate because it implies additivity. When we speak in probabilistic terms we have implicitly in mind a complete set of events that will take place and we assume that one of them will certainly take place. But we do not know the future events. Fuzzy set theory excludes additivity and this is the reason we suggest it for planning models. The state equation

$$x_{k+1} = x_k + u_k$$

where x and u are fuzzy sets, states the intuitively and widely accepted idea that what we can reasonably assume about the situations or events of any future date becomes vaguer and more precarious as we shift our thoughts to more and more remote dates. Shackle[101] notes that as the vista lengthens, the divergence of possible situations from the known present must widen in such a way that each new discovery and invention opens the door to a hundred others and each technological environment produces unguessable reactions. What the decision-maker may then be able to discern is some boundary limits. He rejects the probability language which inhibits it from allowing for novelty. Probability is a unity to be divided out among the constituent contingencies of the set. When each contingency has been assigned its probability, these probabilities add up to one. If then, we are to discard the notion of a known set of contingencies, it seems that we must exclude any additive character of our chosen uncertainty-language. That language is required to assign standing, that is seriousness or realism of any hypothesis in terms that do not have to add up to a total representing certainty. All experience teaches that knowledge is never complete. But if we doubt that all is known, it is misleading to assign probabilities on the basis that all contingencies can be fully partitioned and completely listed. Insufficiency of knowledge is least dangerous when it is acknowledged. The ground for supposing knowledge insufficient is a part of knowledge, and the insufficiency itself must be taken account of in describing the state of thought and its uncertainty. Non-additivity is more incisive than the piling-up of probabilities of the erroneousness of probabilities.

In the view of Georgescu-Roegen the concepts are not arithmomorphic. In a recent book[3] he notes that quantity cannot be regarded as a notion before quality, either in the logical or evolutionary order. Undoubtedly, he says, before the thought of measuring quantities of wheat man must first have come to recognize that one pile of wheat is greater than another without weighing them. For a long time 'colder' and 'hotter' had no measure. Distinctions between 'earlier' and 'later' which reflect qualitative differences must have long preceded the practice of quantitity measure. Further, he observes that risk describes the situations where the exact outcome is not known but the outcome does not represent a novelty. He notes that uncertainty applies to cases where the reason we cannot predict the outcome is that the same event has never been observed in the past.

It is said that the best books are not the ones that give all the answers, but those that are full of provocative points of view, rich examples, and inconclusive discussions, for to those books you must return again and again, this very inconclusiveness makes them take a new aspect every time you return to them with your own new background, and they will always be generous in providing material for new considerations. Among the theses that Georgescu-Roegen[3] defends is the claim that the undeniably difficult problem of describing qualitative change stems from the arithmomorphic schematization. Arithmomorphic concepts do not overlap; every one of them is as discretely distinct as a single number in relation to the infinity of all others. The concepts used in human judgements like 'good' or 'justice' have no arithmomorphic boundaries. Instead they are surrounded by a penumbra within which they overlap with their opposites. At a particular historical moment, he notes, a nation may be both a democracy and a nondemocracy just as there is an age when a man is both young and old. To the category of concepts we cannot apply the fundamental law of logic, the principle of contradiction: '*B* cannot be both *A* and non-*A*'. On the contrary, we must accept that, in certain cases, '*B* is both *A* and non-*A*' is the real statement. Since the latter principle is a cornerstone of dialectics, Georgescu-Roegen refers to the concepts that may violate the principle of contradiction as dialectical.

Though they are not discretely distinct, dialectical concepts are nevertheless distinct. The difference is this: a penumbra separates a dialectical concept from its opposites. In the case of an arithmomorphic concept the separation consists of a void. The extremely important point is that the separating penumbra itself is a dialectical concept.

Like all inventions, that of the arithmomorphic concept also has its good and its bad features. Although it has speeded the advance of knowledge in the domain of inert matter, no process of change can be explained by it. Human propensities, which are the main vehicle of economic change are not arithmomorphic concepts. The obvious conclusion is that if economics is to be a science not only of observable quantities but also of man, it must rely extensively on dialectical reasoning which cannot be exact. Perhaps the most obvious merit of an arithmomorphic model is that which is acknowledged by almost every criticism of mathematical economics, the merit of bringing to light important errors in the works of literary economists who reasoned dialectically. The second role of an arithmomorphic model circumscribes the *raison d'être* of what currently passes for economic theory, which is to supply our dialectical reasoning with a firm backbone. Further, Georgescu-Roegen notes that the essence of development consists of the organizational and flexible power to create new processes rather than the power to produce commodities by materially crystallized plants. As far as one may search the economic literature, all dynamic models, including those concerned with growth, allow for the production of commodities but not for that of processes. The omission is not inconsequential, be it for the theoretical understanding of the economic process or for the relevance of these models as guides for economic planning. A dynamic model is useless for throwing any light on the problem of how growth comes about, which includes the problem of how growth itself may grow faster!

Chapter 5

THE CATEGORICAL APPROACH

5.1. A CATEGORY OF FUZZY SETS

This section must be viewed as an attempt to study social phenomena by means of the theory of fuzzy sets.[102] For the purpose of discussion, let us understand social science in the broadest sense, including sociology, economics, political science, etc. According to the dictionary, *social* means relations to human society, and *social science* is the study of human society, particularly of its organization and of the relationship to it of individual members. Therefore, in the following, we consider the study of human groupings, i.e. any number of people associated geographically, racially or otherwise, from small groups like the family or working crew, to the innumerable intermediates of informal and formal organizations, and on to the largest units such as nations or power blocks. Conceptual models which in simplified and therefore comprehensible form try to represent certain aspects of reality are basic in any attempt at theory. Social scientists who deal with systems involving human components do particularly need analytical frameworks which, rather than obscuring the behavioural implications of the systems, explicitly consider these human factors.

In common with fuzzy sets theory, we suggest new foundations for the theoretical study by constructing fuzzy models, i.e. abstractions of the real world in which the relevant relations between the real elements are replaced by similar relations between suitable mathematical entities. We say suitable, because the many attempts to provide theoretical formulations are all based on a widely held belief that the conventional quantitative techniques of systems analysis successfully used in unravelling the secrets of nature are suitable for dealing with social systems. The fact is that social processes are generally characterized by a great complexity, mainly due to a powerful interaction among their elements which prevents a meaningful and complete description of the system itself, considered in its wholeness. An exact description of any real social phenomenon is virtually impossible. It could be thought then that the greater simplicity of a model may make it easier to grasp the nature of the phenomenon in question. But to make it conceptually controllable, we have to reduce reality to a skeleton. To quote Bertalanffy: 'In doing so, the danger is to cut out vital parts of the anatomy. The danger of oversimplification is greater the more multifarious and complex the phenomenon. This applies not only to grand theories of culture and history but to models we find in any psychological and sociological journal'.

The point we wish to make is that this constitutes the concept of a set proven to be the basis for a good number of mathematical theories used to address problems of contemporary interest and very unsuitable for the type of problem we normally face.

The need to use fuzzy models can be easily seen by considering some examples. In social mobility, for instance, a statement such as 'a son follows his father' implies from a set theoretical point of view that there is a well-defined set of followers, and that a certain individual is identified as a member of that set. But, it is difficult to give a sensible definition of something of that kind. The main problem is that we find a definite, but highly artificial boundary between son in followers and non-followers. There are always sons who are not precisely followers, nor exactly non-followers.

In the same way the diffusion of news between different groups depends, to a large extent, on the relations between groups. Either because these relations are not fully specified or because of the vague character of human behaviour, there is a broad spectrum of links between social groups.

This is why considerable attention has been focused in the last few years on the problem of sets that have imprecise boundaries. A natural, as well as general framework for dealing with such problems was constructed around the notion of a fuzzy set that extends the concept of membership in a set to situations in which there are a continuum of membership.

In spite of the fact that fuzzy means 'blurred', 'indistinct' and 'not clear', a mathematically precise framework is developed in which to treat systems or phenomena that (due to intrinsic fuzziness as distinguished from mere statistical variation) cannot be precisely characterized.

A fuzzy set is an assignment of a value in a lattice to each point in an underlying set X. Thus, if the underlying set is X and the lattice is L, a fuzzy set is a function $F: X \rightarrow L$. In other words, the concept of a fuzzy set could be viewed as a generalization of the concept of preferred alternative, where 'is preferred to' is generalized to 'has a higher grade of membership than'. Therefore, the concept of a subset, i.e. a map $X \rightarrow \{0, 1\}$ can be extended naturally to fuzzy subsets, i.e. maps $X \rightarrow [0, 1]$, because the unit interval is a lattice.

The main idea is that in some circumstances it is insufficient to define a property in such a way that an object either has a property or does not. Rather, it is often more convenient to let objects have some properties to a certain degree. The idea that an object may have a property to a degree rather than on a binary basis is especially important in dealing with subjects related to man and his environment.

The theory of fuzzy sets is rich enough in operations and properties to be of genuine use in constructing models for a wide variety of situations. Take, for instance, the situation of social stratification. By social status we shall not understand a division of society but rather a gradation of positions determined by a variety of factors including, but not limited to, an economic one.

In the classification of individuals, subjective information plays an important role. This subjective information may be represented by a fuzzy relation that corresponds to the subjective connectivity.

A relation is the way in which one thing is associated or connected with another. The intuitive concept of relation has been given a precise meaning in modern mathematics. A relation is a subset R of the Cartesian product $X \times Y$ of two spaces X and Y. A fuzzy relation from X to Y is a fuzzy subset of $X \times Y$ characterized by a membership function F_R which associates with each pair (x, y) its grade of membership $F_R(x, y)$ in R, or the strength of the relation between x and y.

A fuzzy partial ordering, P, is a fuzzy relation which is reflexive, antisymmetric and transitive; that is

$$F_p(x, x) = 1$$
$$F_p(x, y) > 0 \quad \text{and} \quad x \neq y \Rightarrow F_p(y, x) = 0$$
$$F_p(x, z) \geqslant \bigvee_y \, [F_p(x, y) \wedge F_p(y, z)]$$

These orderings frequently occur in utility theory. A similarity relation S is a fuzzy relation which is reflexive, symmetric and transitive, i.e.

$$F_s(x, x) = 1$$
$$F_s(x, y) = F_s(y, x)$$
$$F_s(x, z) \geqslant \bigvee_y [F_s(x, y) \wedge F_s(y, z)]$$

The transitivity is defined here as 'the best wrong way'. Thus, a similarity relation is a fuzzy equivalence relation. In other words, the similitude induces an equivalence relation; consequently we are able to classify. The basic classification problem consists in grouping a set of individuals by assigning them to a number of subpopulations in such a way that the members within each subpopulation are sufficiently alike to justify ignoring the individual differences between them.

In general, an exact formulation of the classification problem must be made in terms of obtaining natural and homogenous groupings. A good approach to the solution of this problem is provided by clustering techniques which consist of separating fuzzy subsets in such a way that each individual resembles more the objects within its fuzzy set than objects outside it.

We shall now turn to find ways of looking at thinking and talking about the complex sets of interactions that have become the major characteristics of contemporary problems. We know that just about everything relates to just about everything else and we want to model these relationships.

Bearing in mind that facts about the behaviour of a social subject can be deduced from the behaviour of the model and that the accuracy of these deductions will depend on the success with which the model embodies the features of the human behaviour, we can go further to characterize not only the class of social groupings viewed as fuzzy sets, but the category of these fuzzy sets that also includes the morphisms or transformations of these groupings. Being highly abstract, category theory is suitable for eliminating distracting detail from highly structured situations. In fact, in any social science one tends to study phenomena through universal properties that characterize them, rather than through description of their anatomy. In other words, one tends to see the forest and not only the trees.

Thus, we can define the category SG of social groupings, which has social groupings as objects and interactions of one groupings to another, as morphisms. Clearly, there is an identity interaction of any grouping by itself, and obviously we can compose two interactions to obtain a third composite interaction. Moreover, this operation is evidently associative.

Let $A:X \to L$; $B:X \to L$ be two social groupings, L being a completely distributive lattice. We shall define $A \times B:X \times Y \to L \times L$ as $(A \times B)(x, y) = (A(x), B(y))$. If $\wedge : L \times L \to L$ is a structure operator of the lattice L (i.e. $a \wedge b = \inf(a, b)$, $a, b \in L$) then $A \otimes B = \wedge \circ (A \times B):X \times Y \to L$ is the Cartesian product of the two fuzzy sets A, B. Since $F_L(x)$ denotes the set of all fuzzy sets of X, then $A \otimes B \in F_L(X \times Y)$. Our intention is to organize the collection of all social groupings as a category SG(L). The objects of this category will be fuzzy sets $A:X \to L$; for two objects $A, B \in |\mathrm{SG}(L)|$ we shall define the morphisms $f:A \sim B$ as elements of $F_L(X \times Y)$ such that

$$f(x, y) \leqslant (A \otimes B)(x, y) = A(x) \wedge B(y), \ \forall (x, y) \in X \times Y.$$

If $C:Z \to L$ is a third fuzzy set and $g:B \sim C$ is a fuzzy morphism, we shall define the composition $g \circ f:A \sim C$

$$\text{by } (g \circ f)(x, z) = \bigvee_{y \in Y} (g(y, z) \wedge f(x, y))$$

We can see that the composition of two fuzzy morphisms leads to a fuzzy morphism, that is

$$(g \circ f)(x, z) \leqslant (A \otimes C)(x, z) = A(x) \wedge C(z)$$

It is easy to see that

$$(g \circ f)(x, z) = \bigvee_{y \in Y} (g(y, z) \wedge f(r, y)) \leqslant \bigvee_{y \in Y} (B(y) \wedge C(z)) \wedge (A(x) \wedge B(y))$$

Using the indempotency property

$$\bigvee_{y \in Y} (A(x) \wedge B(y) \wedge C(z) = A(x) \wedge C(z) \wedge \left[\bigvee_{Y} B(y) \right] \leqslant A(x) \wedge C(z)$$

the associativity is easily verified:

$$(f \circ g) \circ h = f \circ (g \circ h)$$

We shall show that there is a morphism $id_A \in SG(L)(A, A)$ which acts as left and right identity for two elements of $SG(L)(A, B)$ and $SG(L)(B, A)$ respectively, for all objects $B \in |SG(L)|$

We shall define $\Delta_A : A \quad A$ as

$$\Delta_A(x, x') = \begin{cases} A(x), & x = x' \\ 0 & , x \neq x' \end{cases}$$

Δ_A is a morphism since $\Delta_A(x, x') \leqslant A(x) \wedge A(x')$

If $f: A \quad B$, then $f \circ \Delta_A = f$ because

$$(f \circ \Delta_A)(x, y) = \bigvee_{x' \in X} (f(x', y) \wedge \Delta_A(x, x') = f(x, y)$$

$$\forall (x, y) \in X \times Y$$

Similarly, $\Delta_B \circ f = f$

We do not need precise concepts of social grouping and interconnection to consider SG. On the contrary, we wish to deduce things about the structure of representation at the end of the argument. Stated another way, a categorical framework can provide a powerful guide to research directions.

5.2. FUZZY SETS AS FAMILIES OF SETS

Since there are contexts in which it is important to be as precise as possible, it is worth considering the extent to which we can be rid of vagueness. It seems that vagueness can be removed by quantification. For instance, we may redefine a bush to mean a tree having a height of at most 1 metre. But it is by no means the case that we have rendered the term with perfect precision. 'Bush' is still subject to any vagueness attaching to tree. Any project to rid our language of vagueness is chimerical (Alston[103]). For all practical purposes we must use words to provide more precise definitions of other words, and the task of removing the fresh elements of vagueness that come to light with new definitions seems to be endless.

It is an essential characteristic of a vague predicate that the boundaries of the domain of its applicability are not fixed and, therefore, we do not know precisely where this domain ends and some other begins. The question of truth and falsity here is not only undecided but undecidable. And if the lack of a criterion for deciding truth or falsity of a proposition is implied by the very nature of that proposition, it shows the inapplicability of the notions of truth and falsity used in the principle of the excluded middle (Verma[104]). One may object here, and say that the principle of the excluded middle can be made applicable by making the vague concept precise; which may take place in different ways. One method may be to effect an overall change in the system of concepts and to make a completly fresh classification in order to account for the boundary cases. Against this one may argue that vagueness can never be completely removed because there will always be border cases. The objection against the defence of the principle of excluded middle should not be that vagueness cannot be eliminated, but rather that the removal of vagueness does not show that the principle of the excluded middle is applicable to those propositions in which vague predicates occur. By making a vague predicate precise we substitute one predicate for the other, and the proposition in which the precise concept occurs is different from the one in which the vague concept occurs.

What is shown is that vague concepts can be given up and precise ones may be substituted in their place. But, after all, the fact remains that the principle of the excluded middle does not apply to the propositions containing vague predicates. The contention that the principle of the excluded middle is inapplicable to some cases is not contradicted by just substituting for those cases some others to which it is applicable. Making a thing precise removes the reason for the principle being inapplicable, but this does not show it is never a reason. And the fact remains that wherever vagueness is present the principle of the excluded middle will remain inapplicable. In what follows we shall see how we can make it applicable.

In order to obtain a more precise description of the inexactness we encounter in real life, several policies have been proposed. Zadeh[6] introduced the concept of fuzzy sets by defining them in terms of mappings from a set into the unit interval. Goguen[105] extended the concept by defining L-fuzzy sets to be functions from a set into a lattice. Gentilhomme[106] introduced the concept of *flou sets* by defining them as pairs of sets.

Here we shall define the concept of L-flou set in order to show the equivalence with L-fuzzy sets in the sense of a lattice dual isomorphism, L being a complete lattice. These results lead to the following conclusions: (*a*) A fuzzy set may be represented as an infinite family of sets rather as a function; (*b*) a flou set can be reduced to an L-fuzzy set.

As practical applications of these results, we shall prove that fuzzy systems can be represented as families of deterministic systems. The representation theorem will also be used to solve a fuzzy linear programming problem.

Let X be a reference set and L a lattice.

An *L-fuzzy subset of X* (or, shortly, L-set) is a function $f: X - L$. We shall denote by

$$\mathscr{F}_L(X) = \{f / f: X \to L\}$$

the set of all L-fuzzy subsets of X. It is clear that $\mathscr{F}_L(X)$ is also a lattice, as L.

If $L = [0, 1]$ is the unit interval of reals, we shall denote $\mathscr{F}_{[0,1]}(X)$, and call an $f \in \mathscr{F}(X)$, a *fuzzy subset* of X. Generally speaking, $\mathscr{F}_L(X)$ has the same structure as L, for example:

L complete lattice $\Rightarrow \mathscr{F}_L(X)$ complete lattice
L Boolean algebra $\Rightarrow \mathscr{F}_L(X)$ Boolean algebra

Let X now be a set, and L a complete lattice.

An *L-flou subset* of X (or shortly, L-flou set) is a function $u: L \to \mathscr{P}(X)$, with the properties

$FL1)$ $u(\bigvee_{i \in I} \alpha_i) = \bigcap_{i \in I} u(\alpha_i), (\forall)(\alpha_i)_i \subset L$

$FL2)$ $u(0) = X$

This concept generalizes flou sets ('ensembles flous') which appeared first at Gentilhomme. More precisely, a *flou subset* of X is a pair (A, B), where $A \subseteq X$, $B \subseteq X$, and $A \subseteq B$.

If we consider the lattice $L = \{0, \frac{1}{2}, 1\}$, we may see that an L-flou set can be identified with a flou set. In the same way, if $L = \{0, \frac{1}{n}, \frac{2}{n}, \ldots, n - \frac{1}{n}, 1\}$, then an L-flou set can be thought as a family $(A_1, A_2, \ldots, A_n)$, with $A_1 \subseteq A_2 \subseteq \ldots \subseteq A_n$, and $A_i \subseteq X$ for each $i = 1, 2, \ldots, n$. The set of all L-flou subsets of X will be denoted by $\mathscr{F}l_L(X)$.

More particularly, if $L = \{0, \frac{1}{2}, 1\}$ as in above example, we shall denote $\mathscr{F}l_L(X)$ by $\mathscr{F}l(X)$. Our main aim is to prove that L-fuzzy sets and L-flou sets are, in a sense, isomorphic. Then, we shall use this result in order to represent fuzzy systems.

We mention here, however, that almost all 'fuzzy structures' that have appeared in the literature, can be represented as families of classical structures (for more details, *see* Negoita and Ralescu[49]).

As Goguen pointed out,[99] fuzzy sets can be organized as a category. If L is a fixed lattice, the category of fuzzy sets will be denoted by Set (L). The objects of Set (L) are pairs (X, f), where X is a set, and $f: X \to L$. A morphism between two objects (X, f), (Y, g) will be a function $\alpha: X \to Y$, such that in the diagram

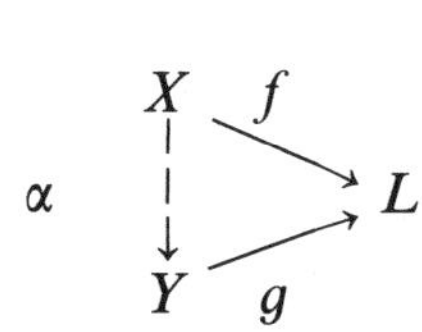

we have $g \circ \alpha \geqslant f$.

It is a simple matter now to define fuzzy systems, according to the definition of dynamical systems in a category (*see* Arbib and Manes[107]).

First we shall define deterministic systems by a complex

$$\mathcal{S} = \{X, U, Y, \delta, \beta, x_0\}$$

where X, U, Y are sets (states, inputs, and outputs, respectively), and $(\delta : X \times U \to X)$ (the dynamics), $(\beta : X \to Y)$ (the output map), and $(x_0 \in X)$ (the initial state). Briefly speaking, a fuzzy system will have as state, input, and output 'objects', fuzzy sets, respectively.

More precisely, a *fuzzy system* is a complex

$$\mathcal{S}_f = \{(X, f), (U, g), (Y, h), \delta, \beta\}$$

where (X, f), (U, g), (f, h) are objects of Set (L), while δ, β are morphisms in Set (L)

$$\delta : (X, f) \times (U, g) \to (X, f)$$
$$\beta : (X, f) \to (Y, g)$$

We mention that category Set (L) has direct products, and
$(X, f) \times (U, g) = (X \times U, f * g)$,
where $f * g : X \times U \to L$, $(f * g)(x, u) = f(x) \wedge g(u)$

We shall focus our attention now on the topic entitled *representation of fuzzy sets*. Our main result in this section is to prove that L-fuzzy sets and L-flou sets are dually isomorphic. We shall consider a (fixed) complete lattice L; thus L has 0 and 1.

Theorem 1 (of representation). There exists a dual isomorphism of complete lattices, between $\mathcal{F}_L(X)$ and $\mathcal{F}l_L(X)$.

Proof. We must define two maps

$$\mathcal{F}_L(X) \underset{\Psi}{\overset{\Phi}{\leftrightarrows}} \mathcal{F}l_L(X)$$

such that

(*a*) $\Phi \circ \Psi = id$, $\Psi \circ \Phi = id$

(*b*) $\Phi(\bigvee_{i \in I} f_i) = \bigwedge_{i \in I} \Phi(f_i)$, $\Phi(\bigwedge_{i \in I} f_i) = \bigvee_{i \in I} \Phi(f_i)$

Let us set $u_f(\alpha) = \{x \in X / f(x) \geqslant \alpha\}$, $(\forall)\alpha \in L$.

Since $\{x \in X / f(x) \geqslant \bigvee_{i \in I} \alpha_i\} = \bigcap_{i \in I} \{x \in X / f(x) \geqslant \alpha_i\}$, and $\{x \in X / f(x) \geqslant 0\} = X$, the result is that Φ is well defined.

Conversely, we set $\Psi(u) = f_u$, where $x \in X \Rightarrow f_u(x) = \bigvee_{x \in u(\alpha)} \alpha$

The key point of the proof will be the equalities in (*a*) above. We have

$$(\Phi \circ \Psi)(u)(\alpha) = \Phi(\Psi(u))(\alpha) = \Phi(f_u)(\alpha) = \{x \in X / f_u(x) \geqslant \alpha\}$$
$$= \{x \in X / \bigvee_{x \in u(\beta)} \beta \geqslant \alpha\} = u(\alpha)$$

The last equality results from the fact that $u \in \mathscr{F}l_L(X)$ implies $\alpha \leqslant \beta \Rightarrow u(\alpha) \supseteq u(\beta)$, and, of course, $u(\bigvee_{i \in I} \alpha_i) = \bigcap_{i \in I} u(\alpha_i)$.

Thus $\Phi \circ \Psi = id$.

Conversely, we have

$$(\Psi \circ \Phi)(f)(x) = \Psi(\Phi(f))(x) = \Psi(u_f)(x) = \bigvee_{x \in u_f(\alpha)} \alpha$$

But $\bigvee_{x \in u_f(\alpha)} \alpha = \sup \{\alpha / f(x) \geqslant \alpha\} = f(x)$, thus $\Psi \circ \Phi = id$.

It is now a simple matter of calculus to obtain the equalities (*b*), which finally prove that $\mathscr{F}_L(X)$ and $\mathscr{F}l_L(X)$ are dually isomorphic (as complete lattices).

It is clear that an *L*-flou set $u \in \mathscr{F}l_L(X)$, can be thought of as a family $u = (X_\alpha)_{\alpha \in L}$, $X_\alpha \subseteq X$, with properties

*FL*1) $X_{\bigvee_{i \in I} \alpha_i} = \bigcap_{i \in I} X_{\alpha_i}$, $(\forall)(\alpha_i)_{i \in I} \subset L$

*FL*2) $X_0 = X$

Thus, Theorem 1 states that *L*-flou sets can be identified with such families of ordinary sets.

(*a*) If we take $L = [0, 1]$, we obtain the result: fuzzy sets are equivalent with families $(X_\alpha)_{\alpha \in [0, 1]}$ with properties (*FL*1) and (*FL*2) above.

(*b*) If $L = \{0, \frac{1}{2}, 1\}$, we obtain that flou sets can be represented by functions $f: X \to \{0, \frac{1}{2}, 1\}$, thus by *L*-fuzzy sets. This representation theorem can be used to derive representations for such fuzzy concepts as fuzzy convex sets, similarity relations, fuzzy orderings and fuzzy groups.

The idea is to replace in a problem a fuzzy concept by a family of classical concepts, and then to apply the classical theory. This can lead to the possibility of solving many 'fuzzy problems' by reducing them to classical problems. We shall give in the following two examples of using the above representation theorem—representation of fuzzy systems, and the study of a fuzzy linear programming problems.

(i) *Representation of fuzzy systems*

We shall consider, in the following, a state space *X*, an input space *U*, and an output space *Y*.

Let also $\mathscr{S} = \{X, U, Y, \delta, \beta\}$ be a dynamical system. We say that $\mathscr{S}_0$ is a *subsystem* of $\mathscr{S}$, denoted by $\mathscr{S}_0 \subseteq \mathscr{S}$, if $\mathscr{S}_0 = \{X_0, U_0, Y_0, \delta_0, \beta_0\}$ and

(*a*) $X_0 \subseteq X, U_0 \subseteq U, Y_0 \subseteq Y$
(*b*) $\delta / X_0 \times U_0 = \delta_0, \beta / X_0 = \beta_0$

A *fuzzy subsystem* of $\mathscr{F}$, is a fuzzy system

$$\mathscr{F}_f = \{(X, f), (U, g), (Y, h), \delta, \beta\}$$

If we consider a fixed system $\mathscr{S}$, we shall denote the set of all its fuzzy subsystems by $\mathscr{F}(\mathscr{S})$.

We shall also denote the set of all families $(\mathscr{S}_\alpha)_{\alpha \in L}$ by $\mathscr{F}l(\mathscr{S})$ where $\mathscr{S}_\alpha$ is a subsystem of $\mathscr{F}$, for each $\alpha \in L$ such that:

(*a*) $\mathscr{S}_{\bigvee_{i \in I} \alpha_i} = \bigcap_{i \in I} \mathscr{S}_{\alpha_i}, (\forall)(\alpha_i)_{i \in I} \subset L$

(*b*) $\mathscr{S}_0 = \mathscr{S}$

L is a (fixed) complete lattice. In the definition of $\mathscr{F}l(\mathscr{S})$ we have also used the *intersection* of subsystems, which is defined in an obvious way.

Theorem 2. There is a bijection between the sets $\mathscr{F}(\mathscr{S})$ and $\mathscr{F}l(\mathscr{S})$.

Proof. We must build two functions

$$\mathscr{F}(\mathscr{S}) \underset{\Psi}{\overset{\Phi}{\leftrightarrows}} \mathscr{F}l(\mathscr{S})$$

such that $\Phi \circ \Psi = id, \Psi \circ \Phi = id.$

Let $\mathscr{S}_f = \{(X, f), (U, g), (Y, h), \delta, \beta\} \in \mathscr{F}(\mathscr{S})$.
For a fuzzy set $f: X \to L$, we shall denote by $L_\alpha(f) = \{x \in X / f(x) \geqslant \alpha\}$, $\alpha \in L$, its α-level sets.

The function Φ is defined by:

$$\Phi(\mathscr{S}_f) = (\mathscr{S}_\alpha)_\alpha, \mathscr{S}_\alpha = \{L_\alpha(f), L_\alpha(g), L_\alpha(h), \delta_\alpha, \beta_\alpha\}$$

In order to avoid complicated notations, it is understood that δ_α in $\mathscr{S}_\alpha$ means the restriction of δ(in $\mathscr{S}_f$) at $L_\alpha(f) \times L_\alpha(g)$.

We must prove, first, that $\mathscr{S}_\alpha$ is well defined, i.e. that

$$\delta_\alpha : L_\alpha(f) \times L_\alpha(g) \to L_\alpha(f), \beta_\alpha : L_\alpha(f) \to L_\alpha(h).$$

But $\delta : (X, f) \times (U, g) \to (X, f)$ is a morphism in the category Set (L). Thus we have: $x \in X, u \in U \Rightarrow f(\delta(x, u)) \geqslant f(x) \wedge g(u)$.

If $x \in L_\alpha(f), u \in L_\alpha(g) \Rightarrow f(x) \geqslant \alpha, g(u) \geqslant \alpha$, thus

$$f(\delta(x, u)) \geqslant \alpha \Rightarrow \delta(x, u) \in L_\alpha(f)$$

It is simple to prove, finally, that $(\mathscr{S}_\alpha)_{\alpha \in L} \in \mathscr{F}l(\mathscr{S})$. Conversely, let $(\mathscr{S}_\alpha)_{\alpha \in L} \in \mathscr{F}l(\mathscr{S})$ be a family of subsystems of $\mathscr{S}$, with properties (a) and (b) above.

Let $\mathscr{S}_\alpha = \{X_\alpha, U_\alpha, Y_\alpha, \delta_\alpha, \beta_\alpha\}$, $\alpha \in L$. We set

$$\Psi((\mathscr{S}_\alpha)_{\alpha \in L}) = \mathscr{S}_f = \{(X, f), (U, g), (Y, h), \delta, \beta\},$$

where $f(x) = \sup\{\alpha \in L/x \in X_\alpha\}$, $g(u) = \sup\{\alpha \in L/u \in U_\alpha\}$; $h(y) = \sup\{\alpha \in L/y \in Y_\alpha\}$.

An important step is to prove that, with these definitions, δ and β are morphisms in Set (L) (i.e. $\mathscr{S}_f$ is a fuzzy system).

Let us prove, for example, that

$$f(\delta(x, u)) \geqslant f(x) \wedge g(u), (\forall) x \in X, u \in U.$$

Thus: $\sup\{\alpha/\delta(x, u) \in X_\alpha\} \geqslant (\sup\{\alpha/x \in X_\alpha\}) \wedge (\sup\{\alpha/u \in U_\alpha\})$.

Let us observe that $\delta_\alpha : X_\alpha \times U_\alpha \to X_\alpha$; thus for a fixed $x \in X$, $u \in U$, we have $\{\alpha/x \in X_\alpha\} \cap \{\alpha/u \in U_\alpha\} \subset \{\alpha/\delta(x, u) \in X_\alpha\}$ and the inequality above follows. In the same way we can prove that β is a morphism in Set (L).

Using theorem 1, we can finally prove that $\Phi \circ \Psi = id$ (identity), and $\Psi \circ \Phi = id$.

(ii) *Robust programming*

A classical linear programming problem can be written as

$$\sup_{x \in M} f(x)$$

where $M = \{x \in \mathbb{R}^n/x_1 a_1 + \ldots + x_n a_n \leqslant b\}, f(x) = \langle x, c \rangle$.
The columns of the matrix A are denoted by $a_1, \ldots, a_n$, such that the constraints can be written as $Ax \leqslant b$. We suppose that $a_1, \ldots, a_n, b \in \mathbb{R}^m$, $c \in \mathbb{R}^n$, and that $\langle , \rangle$ are the inner products in $\mathbb{R}^n$. Which is nothing more, classically, than a linear programming problem in decision theory.

There are, however, some problems in which the constraints are inexact and thus are not well modelled by such sets as M above. We can define a linear programming problem with set-inclusion constraints.

The idea is to choose subsets $A_1, A_2, \ldots, A_n, B$ of $\mathbb{R}^m$, and to replace the deterministic constraints by

$$M' = \{x \in \mathbb{R}^n/x_1 A_1 + \ldots + x_n A_n \subseteq B\}$$

If A, B are subsets of $\mathbb{R}^m$ (which is a vector space), we have used the operations

$$A + B = \{a + b/a \in A, b \in B\}$$
$$\lambda A = \{\lambda a/a \in A\}$$

Such a model is well suited to such problems, when $a_1, \ldots, a_n$ (which can be prices, for example) are known only with some 'tolerances'. The problem becomes $\sup_{x \in M'} f(x) = \sup_{x \in M'} \langle x, c \rangle$.

In the case when $B = \{y \in \mathbb{R}^m/y_i \geqslant b_i, i = 1, \ldots, m\}$, the problem $\sup_{x \in M'} f(x)$ can be reduced to a *classical* linear programming problem.

There may also be cases when coefficients $a_1, \ldots, a_n, b$ are inexactly defined. In such cases *robust programming* may be useful.

We shall replace the constraints set M' by

$$M'' = \{x \in \mathbb{R}^n/x_1 f_1 + \ldots + x_n f_n \leqslant g\}$$

where $f_1, \ldots, f_n, g \in \mathcal{F}(\mathbb{R}^m)$.

The problem will be $\sup_{x \in M''} f(x) = \sup_{x \in M''} \langle x, c \rangle$.

Since $f_1, \ldots, f_n, g$ are fuzzy subsets of $\mathbb{R}^m$, we must explain what do we mean by the 'sum' and 'product' above.

If $u, v \in \mathcal{F}(\mathbb{R}^m)$, we set

$$x \in \mathbb{R}^m \Rightarrow (u + v)(x) = \bigvee_{y+z=x} [u(y) \wedge v(z)]$$

$$x \in \mathbb{R}^m \Rightarrow (\lambda \times u)(x) = \begin{cases} u(\lambda^{-1} \times x), & \textit{if } \lambda \neq 0 \\ 0 & , \textit{if } \lambda = 0 \end{cases}$$

Theorem 3. The following programming problems are equivalent:

(*a*) $\sup_{x \in M''} f(x)$, $M'' = \{x \in \mathbb{R}^n/x_1 f_1 + \ldots + x_n f_n \leqslant g\}$

(*b*) $\sup_{x \in N} f(x)$, $N = \bigcap_{\alpha \in [0, 1]} \{x \in \mathbb{R}^n/x_1 L_\alpha(f_1) + \ldots + x_n L_\alpha(f_n) \subseteq L_\alpha(g)\}$.

Proof. This result is obtained by observing that for two fuzzy subsets $u, v \in \mathcal{F}(\mathbb{R}^m)$, we have

$$L_\alpha(u + v) = L_\alpha(u) + L_\alpha(v)$$
$$L_\alpha(\lambda \times u) = \lambda \times L_\alpha(u)$$

It is also clear that $u \leqslant v \Leftrightarrow L_\alpha(u) \subseteq L_\alpha(v)$, $(\forall)\alpha \in [0, 1]$. Thus, the problem

$$(a) \begin{cases} x_1 f_1 + \ldots + x_n f_n \leqslant g \\ \sup f(x) \end{cases}$$

is equivalent to the problem

$$(b)\begin{cases} x_1 L_\alpha(f) + \ldots + x_n L_\alpha(f_n) \subseteq L_\alpha(g) \\ \alpha \in [0, 1] \\ \sup f(x) \end{cases}$$

The last has set-inclusive constraints.

The application of this result may sometimes be difficult, since there is an infinity of constraints, according to $\alpha \in [0, 1]$. We mention, however, that when g is finite-valued, the number of constraints becomes finite. Applications of fuzzy linear programming in planning problems can be found in Section 7.1.

5.3. FUZZY SETS AND TOPOI

The theory of fuzzy systems has turned out to be a simple and convenient tool for quite a variety of problems. The language of fuzzy sets is natural for situations dealing with vague concepts. For a long time, the basic difficulty was to translate these concepts into precise mathematical statements. When we pass from a vague concept to a precisely formulated one we speak about an explanation of the former. Thus, the mathematical notion of 'fuzzy sets' is an explanation of an ordinary notion such as 'concepts'.

There are severe limitations to our ability to work with fuzzy sets. One difficulty in constructing a theory of fuzzy sets is the lack of a proper tool, and the need to use the theory of crisp sets in order to define fuzzy sets. According to the theory of representation, a fuzzy set is viewed as a family of crisp sets. Such a representation was applied in fuzzy programming and some results are reported in Section 7.1.

In spite of these, however, the algebraic aspects of the theory of representation have received almost no systematic treatment. This section attempts to show that the fuzzy set and its representation with a family of crisp sets can be better generalized in the special framework of topoi. In this way, we can reconsider all known algebraic or topological structures.

A category E is an elementary topos if, and only if, it satisfies the following conditions (Wraith[108])

1. E has finite limits
2. E is Cartesian closed
3. E has a subobject classified

Condition (1) states that E has a terminal object 1, such that for any object X there is exactly one morphism $X \to 1$, and it has also pullbacks. Conditions (2) states that we can define an exponential, and that for any object D, there is a natural isomorphism

$$\text{Hom}\,(D \times A, B) = \text{Hom}\,(D, B^A)$$

In other words, to every morphism $f{:}D \times A \to B$ in E there is a unique morphism $g{:}D \to B^A$ in E such that the following diagram commutes ($e(g \times 1) = f$)

$$\begin{array}{ccc} & B^A \times A \xrightarrow{e} B & \\ g \times 1 & \uparrow \quad \nearrow & f \\ & D \times D & \end{array}$$

where e is called evaluation.

Condition (3) states that in the category E there is an object Ω and a morphism $1 \to \Omega$ from the terminal object such that to every monomorphism $m{:}A \to X$, i.e. to every subobject of X, there exists exactly one morphism $\varphi{:}X \to \Omega$ such that

$$\begin{array}{ccc} A & \to & 1 \\ \downarrow & & \downarrow \\ X & \to & \Omega \end{array}$$

is a pullback diagram.

We shall comment on these definitions following MacLane[109]. It is known from the category theory that we may define a subobject of an object X to be a monomorphism, i.e. a morphism which can always be cancelled on the left. But we know that subobjects can be described by characteristic functions. A characteristic function of a subset $S \subset X$ is a function $\varphi{:}X \to \{0, 1\}$ to the set of two truth values 0 and 1. There is an easy way of using pullback to reconstruct a subset of X from its characteristic function φ. Starting from the special subset $\{0\} \subset \{0, 1\}$ and its inclusion monomorphism $\{0\} \to \{0, 1\}$ called the monomorphism 'true', we can recover the subset S by pullback as in the diagram.

$$\begin{array}{cccl} S & \to & 1 & = \{0\} \\ \downarrow & & \downarrow & \text{true} \\ X & \to & \{0,1\} & \end{array}$$

This explains the definition of the subset classifier. The description of a subset S from its characteristic function in terms of pullbacks is exactly illustrated in the diagram

$$\begin{array}{ccccc} U & \longrightarrow & & & 1 \\ \Big| & & & & \| \\ \Big| & & S & \rightarrow & 1 \\ \downarrow & & \downarrow & & \downarrow \\ X & = & X & \rightarrow & \Omega \end{array}$$

that an element u of X lies in the subobject S with characteristic function φ if, and only if, $\varphi u =$ true. The unique morphism true: $U \to 1 \to \Omega$ is the predicate 'truth for U'. Because predicates are built into the system E the usual logic is said to be internalized in a topos.

Let $E(X, \Omega)$ denote the set of all morphisms $X \to \Omega$ in the topos E. Since Ω is the subobject classifier, the elements of this set correspond exactly to the subobjects of X. Now the set of this subobjects is a lattice—this is partially ordered by the evident inclusion of subobjects—there is a 'greatest lower bound', which is just the intersection of subobjects as found by pullback, and there is a 'least upper bound' found by using images. This lattice is actually *Heyting algebra* which, regarded as a category, has exponents. This category is the one with objects, the elements of the lattice, and a unique morphism $a \to b$ if, and only if, $a \leqslant b$ is the partial order of the lattice. Since the greatest lower bound $\wedge$ in the lattice is the product in the associated category, the exponential associates with each pair of elements, b, c, an element $b \Rightarrow c$ such that for all a, $a \wedge b \leqslant c$ if, and only if, $a \leqslant b \Rightarrow c$. Since $a \wedge b \leqslant c$ if, and only if, the set Hom $(a \wedge b, c)$ is nonempty, this statement in lattice language is exactly the bijection defining exponentials Hom $(W \times X, Y)$ $=$ Hom (W, Y^X), which asserts that the functor $\times\, X$ (product with X) is adjoined to the functor $(\ .\ )^X$(raised to the power X).

In the category set, whose objects are usual sets, the exponential is the usual function-set $Y^X =$ Hom $(X, Y) = \{t \mid t : X \to Y\}$ consisting of all morphisms from X to Y, and the evaluation is that function e which takes a function t and an element x to the value $e(t, x) = t(x)$ of t at x. Then, given any function $f : W \times X \to Y$ of two variables the corresponding function $g : W \to Y^X$ is the function given by $g(w)(x) = f(w, x)$. The correspondence $f \to g$ is the bijection defining exponentials. It is easy to show that if Ω is an object in an elementary topos E, we can construct a new topos E/Ω whose objects are the arrows $X \to \Omega$ and whose morphisms $f:\varphi \to \varphi'$ can be defined as $X \xrightarrow{f} X'$ with $\varphi' \circ f = \varphi$.

Let us consider the diagrams

$$\begin{array}{ccc} X & \longrightarrow & \Omega \\ & \searrow & \\ 1_x & & X \end{array} \qquad\qquad \begin{array}{ccc} X & \longrightarrow & 1 \\ \downarrow & & \downarrow \\ X \times \Omega & \to & \Omega \end{array}$$

Given the monomorphism $(\varphi, 1_x): X \to \Omega \times X$, then $\Phi_{(\varphi, 1_x)} = \Omega \times X \to \Omega$ is the characteristic function of X through $(\varphi, 1_x)$ and according to condition (2) we have $\tilde{\varphi}:\Omega \to \Omega^x$. In this way the set of all fuzzy sets defined on the carrier X was generalized to the object Ω^x in the topos E. We shall note that $\tilde{\varphi} \times O = \phi$, where O is the morphism false: $1 \to \Omega$. This morphism is minimal in Hom $(1, \Omega)$ which is a lattice (Heyting algebra), and ϕ is the minimal subobject in Ω^x. Since Ω is the subobject classifier, the elements of X correspond exactly to the subobjects of X which form a lattice. In other words, ϕ is the minimal subobject of Hom (X, Ω). So far we have established that $\varphi: X \to \Omega$ can be viewed as a point in Ω^x. We shall also note that according to our hypotheses the following diagram commutes

$$\begin{array}{cccc} & \Omega \times \Omega & \xrightarrow{\wedge} & \Omega \\ \tilde{\varphi} \times \tilde{\varphi} & \downarrow & & \downarrow \tilde{\varphi} \\ & \Omega^x \times \Omega^x & \xrightarrow{\wedge} & \Omega^x \end{array}$$

and $\tilde{\varphi} \circ t:1 \to \Omega^x$, where $t:1 \to \Omega$ means true, being maximal in Hom $(1, \Omega)$.

Now we shall denote the morphisms $\tilde{\varphi}$ by $ObFl_E$. The map of objects from E/Ω into Fl_E is injective; if we consider two objects $\varphi: X \to \Omega$ and $\varphi': X' \to \Omega$ with $X \neq X'$ then $(\varphi, 1_x) \neq (\varphi', 1_x)$ and $\tilde{\varphi} \neq \tilde{\varphi}'$. Note that $\tilde{\varphi}$ is a homomorphism of monoids and that $\tilde{\varphi}(\alpha)$ is the α-model of $\tilde{\varphi}$, a level of explanation.

Consider $ObFl_E$ the class of $F:\Omega \to \Omega^x$ such that

(a) $F \circ O = \phi$

(b) $F \circ \wedge = \wedge(F \times F)$

(c) $\bar{F}^{-1}(t) = t \times 1_x: X \to \Omega \times X$

where $\bar{F}: \Omega \times X \to \Omega$ the morphism isomorphic with F. Then, there exists a unique $\varphi: X \to \Omega$ such that $\tilde{\varphi} = F$. If we consider two objects $F: \Omega \to \Omega^x$

and $F':\Omega \to \Omega^x$ having properties (*a*), (*b*), (*c*), then $F \to F'$ will be a morphism $\alpha:\Omega^x \to \Omega^{x'}$ such that there exists a unique $f: X' \to X$ with $\alpha = \Omega^f$. In this way we have defined the category Fl_E.

The contravariant functor $\sim : E/\Omega \to Fl_E$ is an assignment of an object $\tilde{\varphi}:\Omega \to \Omega^x$ to each object $\varphi: X \to \Omega$ and of an morphism $a:\tilde{\varphi}' \to \tilde{\varphi}$ to each morphism $f:\varphi \to \varphi'$. This functor has an inverse $T:Fl_E \to E/\Omega$ defined as $T(F) = F \circ t$, where Ft is from Hom $(1, \Omega^x) \simeq$ Hom $(X \times 1, \Omega) \simeq$ Hom (X, Ω). If $F \overset{\alpha}{\to} F'$ is a morphism $\Omega^x \to \Omega^{x'}$, then according to the definition of α there is a unique $f: X' \to X$. Since T and $\sim$ define an isomorphism of categories between Fl_E and E, then Fl_E is a topos. We say that the objects in Fl_E are nuances or interpretations in E.

The idea of representing a fuzzy set by a family of crisp sets was generalized in the framework of the theory of elementary topoi and, in this way, to study a fuzzy object means to study heyting algebras and their homomorphisms. One can apply the results from the theory of elementary topoi and, thus, the theory of fuzzy sets can be developed in a natural manner, and not by simplistic 'fuzzification' of classical theories.

5.4. C-SETS

In this section the concept of C-set is introduced, where C is a suitable category. Operations with C-sets are investigated and the category of C-sets is defined in order to generalize the lattice of fuzzy sets. Further, the representation of C-sets with a category of functors is presented, thus extending practically the results from section 5.2. Two representation theorems are given, one having an atomistic character and the other having a categorical character.

Intuitively, a fuzzy subset of a set X, $f:X \to L$, is a way of ordering the elements of the set X according to their membership of an imprecise statement about X. This way of ordering is unique for a fixed lattice L, and usually $L = [0, 1]$. For example, we can say that x is better than y if $f(x) \geqslant f(y)$. But we cannot say that x is not as good as y. In order to improve the possibility of comparing points in X, one can introduce the concept of a C-set.

Let L be a lattice. We have already seen that we can define a category denoted also by L. The objects are the elements of L $x, y, z, \ldots \in L$. For $x, y \in L$ the morphisms are

$$\text{Hom}(x, y) = \begin{cases} \leqslant & \text{if } x \leqslant y \\ \phi & \text{otherwise} \end{cases}$$

This means that there exists a unique morphism from x in y if, and only if, $x \leqslant y$.

Now, if C is a category and X a set, we shall call C-subset of X a function $f\colon X \to ObC$, from X into the objects of C. It is quite clear that if $C = L$ an L-subset of X is precisely the fuzzy subset of X. The crucial point here is that in this way we have more than one criterion by which to compare the elements of X. For instance, for two elements $x, y \in X$ the set Hom $(f(x), f(y))$ gives information about the orderings between x and y. If ObC is a set, then C is a small category. In order to avoid mappings between classes we shall consider only small categories in our definition of C-set.

Let C be a small category and X a fixed set. We shall further denote by $F_C(X)$ the set of all C-subsets of X

$$F_C(X) = \{f \mid f\colon X \to ObC\}.$$

Bearing in mind that the set of fuzzy sets is a lattice, we are obliged to define two operations in $F_C(X)$ in order to extend the union and intersection of fuzzy sets. Let us suppose that C has finite products and coproducts. We can define, then, for $f, g \in F_C(X)$

$$(f \cup g)(x) = f(x) \coprod g(x) \qquad \text{union}$$

$$(f \cap g)(x) = f(x) \prod g(x) \qquad \text{intersection}$$

Although many properties of these operations can be proved for C-sets we shall see that this way is unnatural. The main difficulty is that we are not able to define in $F_C(X)$ an ordering. This difficulty stems from the categorical nature of a C-set. Let us go more deeply into this problem. A preordering can be defined in $F_C(X)$ by setting

$$f \prec g \Leftrightarrow \mathrm{Hom}_C\,(f(x), g(x)) \neq 0$$

If we consider also the equivalence

$$f \sim g \Leftrightarrow f \prec g \quad \text{and} \quad g \prec f$$

and denote by $\mathscr{F}_C(X) = F_C(X)/\!\sim$, the quotient space, then an ordering can be defined in $\mathscr{F}_C(X)$ by

$$f \leqslant g \Leftrightarrow f \prec g$$

It is a simple matter of calculus to prove that operations $\cap$ and $\cup$ can be extended to $\mathscr{F}_C(X)$ by

$$\hat{f} \cup \hat{g} = f \,\widehat{\cup}\, g$$

$$\hat{f} \cap \hat{g} = f \widehat{\cap} g.$$

This way seems not to be very attractive, since the elements of $\mathscr{F}_C$ are not familiar.

The idea is to consider $F_C(X)$ as a category. Again, the morphisms of this category will generalize the inclusion of fuzzy subsets. The category of C-subsets of X has as objects function $f: X \to ObC$, and a morphism $m: f \to g$ is a family $m = (m_x)_{x \in X}$ of morphisms in C, $m_x: f(x) \to g(x)$. The composition of morphisms is made component wise. We shall denote this category by the same symbol $F_C(X)$. In the classical case, $C = L$, to give for each $x \in X$ a morphism $m_x: f(x) \to g(x)$ means to say that $f(x) \leqslant g(x)$ for all $x \in X$, that is, $f \leqslant g$. This is precisely the case of inclusion of fuzzy subsets.

Our problem now is to study the relationship between the two categories C and $F_C(X)$. We are looking for the implication: if C has products and coproducts then $F_C(X)$ has products and co-products. This implication is very similar to the one valid in the case of lattices: If L is a lattice then the set of all fuzzy sets is also a lattice.

We shall turn our attention now toward the representation problem. For the beginning we shall look for a kind of level structure.

Consider $P(X)$ the set of all subsets of X, which obviously is a category, $f \in F_C(X)$ and $\alpha \in ObC$, and the α-level set of f as the set

$$N_\alpha(f) = \{x \in X \mid \mathrm{Hom}_C(\alpha, f(x)) \neq \phi\}$$

In the case of fuzzy sets, $C = L$, and we have $N_\alpha(f) = \{x \in X \mid f(x) \geqslant \alpha\}$ It follows that for C-sets, a mapping can be defined

$$N(f): ObC \to P(X),\ N(f)(\alpha) = N_\alpha(f)$$

The mapping $N(f)$ is a contravariant functor $N(f): C \to P(X)$. For each morphism $m: \alpha \to \beta$ we have $N(f)(\alpha) \supseteq N(f)(\beta)$. If C has an initial object O we have

$$N(f)(O) = \{x \in X \mid \mathrm{Hom}_C(O, f(x)) \neq \phi\} = X.$$

Using this model of level functors we shall define C-flou set as a contravariant functor $F: C \to P(X)$. The set of all flou subsets of X will be denoted by $Fl_C(X)$

$$Fl_C(X) = \{F \mid F: C \to P(X), \text{ contravariant}\}$$

It is easy to see that Fl_C is a category, whose morphisms are natural transformations between functors.

The question now is in what way C-sets can be represented by flou sets. Note that in this case we no longer have an isomorphism theorem as we had for fuzzy sets, due to the richness of morphisms in the category C. We have here a lot of possibilities to compare with the elements of X.

If the category C has arbitrary products, and for each α, $(\alpha_i) \leqslant ObC$ and for each morphism $\alpha \xrightarrow{m} \coprod_{i\in I} \alpha_i$ there exists $i_0 \in I$ and a morphism $\alpha \xrightarrow{m_0} \alpha_{i_0}$ such that the diagram

$$\begin{array}{ccc} \alpha & \xrightarrow{m} & \coprod_{i\in I} \alpha_i \\ & \searrow_{m_0} & \nearrow_{j_{i_0}} \\ & & \alpha_{i_0} \end{array}$$

commutes, then it is easy to prove Theorem 1 (of representation). There exists a surjection

$$\phi : F_C(X) \to Fl_C(X)$$

Let us also define a function

$$\psi : Fl_C(X) \to F_C(X)$$

The functions Φ and ψ are, in fact, functors which form an adjoined pair. According to Theorem 2 (of representation) the pair of functors satisfy the properties

(1) $\phi \circ \psi = id_{Fl_C(X)}$
(2) there exists a natural transformation
$j : id_{F_C(X)} \to \psi \circ \phi$

Chapter 6

SYNTHESIS PROCESSES

This chapter is an illustration of applicability and explanation of the ideas delineated in the previous chapters. The vaguely defined concept of synthesis presented in chapter 2 is given a meaning through some examples, i.e. particular cases of decision-making. There has been quite a proliferation of methods for modelling decision processes over the years, examples of which include flexible programming and robust programming. We shall try to see what relation these bear to fuzzy systems theory.

Most of the material in this chapter, in one formulation or another, attracted attention among scientists with a wide variety of points of view. We therefore felt it worthwhile to indicate briefly what the tools of the fuzzy systems theory are and what some of the key results tell us.

The examples presented here are taken from management science or, more precisely, from managerial situations. Over the past ten years there has been far-reaching developments in the techniques of applying the ways of thinking and working commonly used by system theorists to problems confronting decision-makers. An extensive body of theory has been built up with respect to many problems of scheduling, programming and planning. In the absence of analogues for such dependable physical laws as those of Newton, management scientists have had to rely more upon the law of balance. A stock represents a quantity fixed at a point in time. Inventories are typically thought of in this way. The state equation of a production system represented the noticeable differences between the state of the system at different times. The succession of these states implies that there are orderly processes which explain how the system gets from its present state to whereover it is going, i.e. the system is heading somewhere. A performance criterion is also considered in order to keep state and control variables near desired values. However, the question, of what is good remains vague and it is not at all strange that criteria based on different backgrounds can be suggested. Different criteria give different conclusions about goodness. Important efforts to develop new theories for resolving problems with multiple criteria have been noticed recently.

Finding problems is easy; realistically solving them is another matter. In complex systems the determination of the structure of the model, apart from the judgement of the performance criterion, is a serious difficulty. A feeling now exists that new techniques must be found, and the theoretical tools of the fuzzy systems seem to be adequate in this field.

6.1. OPEN SYSTEMS

Recent developments in global simulation models and the application of control theory in economics have brought into question the assump-that quantitative models can represent all the complexities of reality. It is now clear that a successful systems approach depends critically on the potential of modelling methods. Some form of modelling paradigm is necessary for the advance of the subject. In order to utilize system concepts in dealing with complex problems of society, a change in the methodology of science is needed. A thesis advanced in this book is that System Theory is currently reaching the responsive frontier where the new skills of handling inexact concepts will become more critical. Enduring fuzziness is the way the decision-makers use in reasoning about actions in a world of some imprecision. Using fuzziness is the way humans perform for getting robustness. Developments in fuzzy set theory have made it possible to offer a formal treatment of vagueness and conflict resolution. Here it is shown that a hierarchical structure contributes to structural stability and that human systems, which are basically open to structural change can be modelled as fuzzy systems.

The notion of complexity basically refers to cognition processes. Human ability to describe nature depends on the mental organization. Language is considered now to be a system of systems, developed by the human mind over the course of millenia.

The theory of cognition argues that concepts are amalgamations of subconcepts, each of which has different conflicting meanings. Briefly, conceptualization is viewed as a process governed by a dialectical qualitative change. If we explore the mode of cognition as a procedure of describing, we can see that humans cope with complexity by assembling partial descriptions. The formation of new descriptions with permanently widening representation builds a hierarchy of semantic levels. This process forms a paradigm for compact representation. Conceptualization is not concerned with details; it is concerned with getting a feel of the whole. It is perfectly clear that people cope with complexity by being able to build abstract concepts. The entire vocabulary of a language is coherently organized into dynamic, operative units in which the meaning is the observable result of an incursion into a mental movement.

We need enough *esprit de finesse* to interpret nouns as 'progress', adjectives such as 'good', and verbs such as 'grow', which are all vague in the sense that a continuum of situations can be found in which at one end the word is satisfied and at the other it is not satisfied, and yet there is no place in between for a discrete change.

Any vague concept can be represented by a fuzzy set, which is a function f defined on a universe X with values in a lattice L. We write $f: X \to L$. A fuzzy set is a description which consists of two sets, the domain and codomain, together with a way of assigning to each element x of the domain an element $f(x)$ called the grade of membership. This function can be used in inferences that explain the process of abstraction.

From a macroscopic examination of the description of a process we can reconstruct the dynamic that generates it. We denote this dynamic by $d: F(X) \times F(X) \to F(X)$ where $F(X)$ is the set of all fuzzy sets defined on X. The interpretation of these symbols is that two fuzzy sets h_1 and h_2 at level k produce a fuzzy set at level $k + 1$ and this is $d(h_1, h_2)$. In other words, if at successive indices k we can have a description from the set $F(X)$, and, further, if we consider at index k, an old description h_1, and a new description h_2, then the description at index $k + 1$ will be $d(h_1, h_2)$. We say that function d describes the movement in the hierarchy of inexact concepts. Clearly, a fuzzy set contains its description, and this is why we use it as a model for the meaning of a concept that manifests itself only by the dynamics that creates it.

Synthesis can be characterized by increasing the level in the hierarchy, i.e. the increasing elaboration of external description. To understand a description means to explain it. To understand the world means to aggregate additional description. How do we go about incorporating descriptions into a new description which refines them both? The obvious way is to take their pullback.

Chang[110] remarks, for instance, that one major problem in economic forecasting is the existence of a large number of models, naïve, time series, structural models, based on different theories, and intuitive unformulated models or any combination of these. In the last analysis, the forecast result of each model can be represented as a fuzzy set. One way to define a consensus among equally creditable conflicting models is to use the intersection. Rosen[111] also agrees that the philosophy of analysis of complex systems assumes that a comprehensive enough family of models will allow us to reconstruct the original system.

The primary virtue of pullbacks is that they permit a unified view. It is important to point out that the hierarchy is employed in addition to, not instead of, the use of the two descriptions. As we face new knowledge, the use of descriptions must be supplemented by other integrating descriptions which present a sort of continuity property that we call robustness. Forgetting inessential details serves the important purpose of compressing redundant information. The ability to encode information at the concept level has practical importance because it saves memory and because all the elements of a concept may not be explicitely known.

If the Wiener definition of information is true, that it is a name for the context of what is exchanged with the outer world as we adjust to it, then it appears reasonable to assume that conceptualization is linked with strong structural stability. At high levels of abstraction, the effect of partial descriptions will be negligible. In other words, small changes in partial descriptions have no effects upon the synthetic description. The property of structural stability arises precisely from the melting of alternate descriptions. We speak about transformations of fuzzy sets such that if we consider the pairs (X, f), $f: X - L$ as objects and the maps $t: X \rightarrow X$ as morphisms; we can then speak about a category of fuzzy sets, a particularly appropriate investigation tool because it allows us to study relations between and operations upon inexact concepts without committing ourselves about their internal structure.

From the present treatment we see quite clearly how abstraction arises from the interplay of separate descriptions of the states of complex systems, how robustness is, in fact, a corollary of transformations of fuzzy sets, a process named fuzzy system, and how in order to study such a system we can use the language of category theory.

Now, it is easy to see that, if gathering partial descriptions is an indication of positive time, then fuzziness increase is perpendicular to the flow of registered items. This seems to be the Bergsonian time, that inner duration, perceived by consciousness, which is nothing else but the melting of states of consciousness into one another and the gradual growth of the ego. It can be said that the learning process is one that proceeds to alter the synthesis, and the direction of development is associated with the rise of synthesis.

As an example of learning system let us consider motivation. Human needs may be organized in a hierarchy starting with basic requirements and proceeding through a series of levels. I have found it more natural and convenient to construct a motivation theory within the framework of fuzzy set algebra. What makes needs inexact concepts is that they change with time and place. Needs in our society today are subjective evaluations. Satisfaction is culturally defined and different from that of two hundred years ago. The material significance underlying each need surely must consist of more fundamental units of meaning, which do doubt are arranged in some coherent order. The only likely manner of engendering the notion of safety from that of satisfaction, for example, appears to be through the addition of a number of elements of meaning. The concept of satisfaction is a structured whole, consisting of a myriad of subsystems. This system, clearly open-ended and in a state of flux, provided that significant materials are constantly being added to cover new aspects of our experience, can be represented as a fuzzy system. That satisfaction is not a direct supporter of experience, but a refinement, a generalization of it shown by the fact that we may represent many needs that are not in direct relationship with the observable world, i.e. abstract needs. At progressively higher levels are aggregations of needs; the displacement of ideal means, in fact, such an aggregation.

Economic and social systems are structurally open and context-sensitive, that is, their basic relationships depend substantially on their particular environment. In my view, the morphogenic tendencies of such systems can be described in terms of fuzzy systems. They are structurally open in order to be structurally stable; more precisely, in order to be robust. The crucial aspect that makes a development process, a learning and not an adaptive process, is the role of synthesis, the mode of melting new knowledge with old experience. An essential feature of learning is the irreversibility of the process. Therefore, if we see a development process as an autopoietic organization which is realized as an autonomous and self-maintaining unity through a network of syntheses-producing processes such that the components, through their interaction generate recursively the same network of processes that produced them, then a fuzzy system is a good model for organic growth.

If artificial intelligence can be viewed as experimental philosophy, then fuzzy mathematics can be understood as a kind of experimental dialectics. Fuzzy mathematics can help us to understand the way in which phenomena have, and therefore will, interact and develop, the general laws of motion that govern the evolution of society, how new syntheses, that preserve the best of opposites arise, and how the processes repeat themselves. By such process is society to be understood.

The Greek 'paradeiknyo' means to show side by side. Mathematics is an old paradigm studying the relationships between magnitudes as represented by numbers. Fuzzy mathematics can be a new paradigm studying the relationships between inexact concepts as represented by fuzzy sets that are functions. It is hoped that fuzzy mathematics can help us to understand spontaneous order and the conditions under which such an order will form itself, the action between opposites out of which the new or higher synthesis emerges.

A social science that hopes to be comprehensive will have to account for the ability of the human being to group individual significant materials into larger syntheses; it cannot ignore learning simply because it is the aspect of morphogenesis that has the closest connections with the external universe. This section has shown that the aim of learning is structural stability and that a possible model of development could be the fuzzy system. Now let us see some examples.

6.2. PULLBACK AND SYNTHESIS

The term 'pullback' is taken from category theory, the intention of which is to establish sets in terms of functions and their compositions. A category is a class of objects together with a class of morphisms between objects. These morphisms may be composed under certain conditions.

To every pair $X'' \xrightarrow{a} X \xleftarrow{b} X'$ of morphisms a and b with common co-domain X, the category contains a pullback (P, p, q). A pullback consists of a new vertex P and morphisms p, q such that the square on P commutes ($ap = bq$ as below) with the following universal property: given any commutative square ($bk = ah$) on the edges a and b with new corner Z there is a unique morphism $s: Z \to P$ such that the whole diagram commutes ($ps = h$, $qs = k$).

In the category set, where objects are all sets, and morphisms are functions, this pullback always exists. There are several useful special

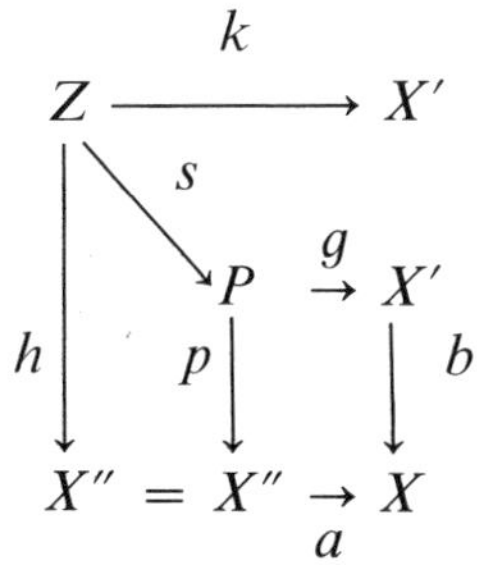

cases. If X'' is the terminal object 1 (such that for any object A there is exactly one morphism $A \to 1$), the square automatically commutes and P is exactly the Cartesian product $X'' \times X'$ is a subset $X' \subset X$ and the function b is the inclusion, then P is that subset of X which is the inverse image of X' under the map $a: X'' \to X$.

Finally, if X'' and X' are both subsets of X and a, b both inclusion maps, then P is clearly just the intersection $X'' \cap X'$. Pullbacks exist in all the categories of fuzzy sets and in each such case they yield products of arbitrary objects, intersection of subobjects and other useful constructions just as in the category set.

Management science deals with complex systems, and can be described in many distinct ways. A basic question then is how do such distinct descriptions of the same system relate to one another. To investigate this question we regard a description as a mapping defined on the space of alternatives. Different descriptions place different evaluations on the set of alternatives and, in particular, provide us with different orders. We say that the diagram commutes if the resulting composite of functions along a path is the same whenever the endpoints are the same.

So far we have considered the objects X, L as being sets. Now, if we handle the objects $X' \to L$, $X'' \to L$, $X \to L$ which are fuzzy sets, and X', $X'' \subset X$ we have the diagram

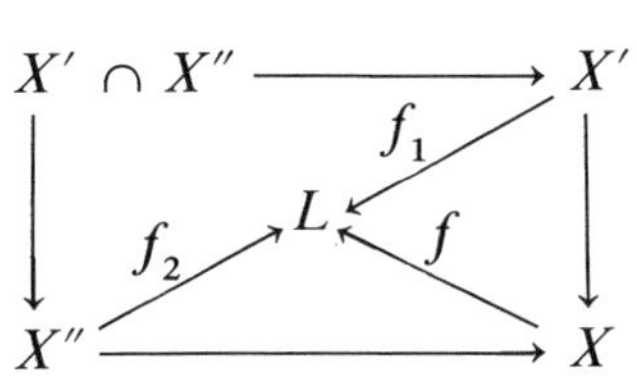

and the pullback $X' \cap X'' \to L$ is given by the diagram

$$\begin{array}{ccc} p & \longrightarrow & f_1 \\ \downarrow & & \downarrow \\ f_2 & \longrightarrow & f \end{array}$$

The primary virtue of pullback is that it permits a unified view; the key in the unification lies in conflict resolution. Inexact descriptions (criteria) are associated with undecidability generated by conflict. Domains of applicability are always overlapping. The existence of conflict at one level of description generates a pullback at the next level. This movement from one level to the other is a transformation of fuzzy sets. The goal of this movement is compact representation. A property of continuity is preserved by melting alternate descriptions. Immense clarity may be offered by modern algebraic theory.

Take, for instance, the case of flexible planning. The role of planning is to show how to make a system behave in some desired way. Most planning models (models of what ought to happen) rely on linear programming techniques. In spite of the extensive work in computational methods for linear programming, little of it has specifically concerned real planning problems. In any linear programming problem (find min $f_0(x)$ subject to $Ax \leqslant b$, $x \in R^n$) the inequalities specify a feasible region. Clearly, inconsistent constraints result in this region being empty. The planner, however, is keen to have an indication on the changes of constraints for acquiring yet a solution. This is why in almost all real planning problems, instead of numbers, planners handle intervals. The real planning problem is to find a number a in an interval $[b, B]$ such that the inequality $Ax \leqslant a$ is a feasible region, and a is as close as possible to b. The meaning of 'as close as possible' is given by the function $F: R \to [0, 1]$ having the support $[b, B]$ such that $a \in [b, B]$ implies $F(a) > 0$.

Bearing in mind that A is a matrix and a a vector, each constraint may be written as a composition $R^n \xrightarrow{A_i} R \xrightarrow{F_i} [0, 1]$. In other words, the diagram

$$\begin{array}{ccc} A_i^{-1}(F_i) & \longrightarrow & F_i \\ \downarrow & & \downarrow \\ R^n & \longrightarrow & R \end{array}$$

commutes. In the language of category theory $A_i^{-1}(F_i) \hookrightarrow R^n$ is the pullback of $F_i \hookrightarrow R$ along A_i. The inverse image of a subobject F_i under a morphism A_i is a subobject $A_i^{-1}(F_i)$ such that the restriction of A_i to $A_i^{-1}(F_i)$ factors through F_i, and if $G_i \to R^n$, $G_i \to F_i$ are such that $G_i \to R^n \to R = G_i \to F_i \hookrightarrow R$, then there is a unique $G_i \to A_i^{-1}(F_i)$ such that the above diagram commutes. This procedure takes maximum advantage of the compromise and hence corresponds closely to the thinking procedure implicitly employed by man. Frontiers of mathematical programming, as they relate to management applications, are pushed forwards if pullback is considered.

Given a set X of alternatives, a set of fuzzy constraints $f_i : X \to [0, 1]$, and a fuzzy goal $f_0 : X \to [0, 1]$, a decision[37] is a fuzzy set $D : X \to [0, 1]$ where $D(x) = \min f_i(x)$, $i = 0, 1, \ldots, m$. The fuzzy programming problem is: find $\bar{x} \in X$ such that $D(\bar{x}) = \sup D(x)$ for all $x \in X$. Now, with this as a background we can motivate and present without proof, a theorem, and illustrate its usefulness with a classical example.[44]

We say that an alternative $x \in X$ is an optimal decision if, and only if, the vector $(\bar{x}, \bar{x}_{n+1}) \in X \times [0, 1]$, where $\bar{x}_{n+1} = \min_{i=0, \cdots, m} f_i(x)$, $\bar{x} = (x_1, x_2, \ldots, x_n)$, is a solution of the following programming problem: find $\max x_{n+1}$ subject to $x_{n+1} < f_i(x)$, $i = 1, \ldots, m$, $x_{n+1} \geqslant 0$.

This is a very useful result, for it says that whenever we have a fuzzy programming problem it requires no extra effort to handle it as a classical one.

Now, we are able to solve a decision problem in a well-known managerial situation. Consider the following scenario, the case of the planning problem fitted into the framework of linear programming. Such a problem is characterized by the constraints $\sum_{j=1}^{n} a_{ij} x_j \leqslant b_i$, $i = 1, \ldots, m$ which specify a feasible region. Facing an empty feasible region, a modification of the constraints seems imperative. The planner is keen to have an indication on the changes of constraints for acquiring a solution yet. In fact, he wants to know the minimum degradation needed in the specifications so that the problem becomes solvable. It is evident that proficiency in changing the coefficients b_i could be a way to find a solution. But how far?

We will suppose that our planner operates in a flexible environment, i.e. instead of numbers he handles intervals. This means that instead of coefficients b_i he uses intervals $[b_i, B_i]$ and all is well if he can choose a number γ in the interval such that $\sum_{j=1}^{n} a_{ij} x_j < \gamma$ is a feasible region.

We shall give a procedure for identifying the vector γ in the interval $[b, B]$ which is as close as possible to b. Our planner's situation can be stated now as: find $\bar{x} = (\bar{x}_1, \bar{x}_2, \ldots, \bar{x}_n) \in R^n$ so that for all $i = 1, \ldots, m$ $\sum_{j=1}^{n} a_{ij}\bar{x}_j \in [b_i, B_i]$, and $\sum_{j=1}^{n} a_{ij}\bar{x}_j - b_i$ is as small as possible. This last constraint may be modelled as a fuzzy set $F_i : R^n \to [0, 1]$.

$$F_i(x) = \frac{B_i - \sum_{j=1}^{n} a_{ij}x_j}{B_i - b_i}$$

Bearing in mind our previous theorem, and with the notation $x_{n+1} = \min F_i(x)$, the planner now handles the following simple problem: Find $\max x_{n+1}$ subject to $\sum_{j=1}^{n} a_{ij}x_j + (B_i - b_i)x_{n+1} < B_i$. The advantage of converting the problem to such a form is evident; it relates to the usual computational procedures used in locating optimal solutions. Clearly, in this case fuzziness is no more a disagreeable feature of the planning problem; it has turned out to be a rather fortunate and appropriate one.

Although sketchy, this section was intended to provide a feeling of how to cope with fuzziness in some managerial situations. The whole idea is that many models—fuzzy in nature—may be translated into deterministic terms. Our second main result is even more surprising; it shows that lack of precision can be assumed for flexibility's sake. The proof techniques used are significant in their own right.

Chapter 7

FUZZINESS IN MANAGEMENT

A silver jubilee is an occasion for looking back and for thinking about the path ahead: 1978 was the 25th year of the existence of the Institute of Management Science (TIMS). At the New York meeting in May 1978, some of the charter members and presidents shared their thoughts on this auspicious occasion. Edward Barankin, for instance, maintained that management science was in trouble. He believed the causes of the trouble lay in grinding out solutions according to concepts and principles prevailing in the fields of economics, statistics, and so on. He was against the naïve mathematics of utility, and of the statistical idea that all the complex attitudes and feelings of a human being on a particular matter can be summed up in a simple numerical loss function. He considered that while their adherents carry on their pontificating in ivory towers, and while the incompetency of their concepts grows more and more evident all round, management science, on the line, meeting face-to-face with its customers, gets the door slammed in its face for that incompetency.

Murray Geisler believed that the most important technical advances in management sciences were being made in the subjective approach to high level decision-making. The ideas of multiple objectives, uncertainty, and fuzziness were he said where the newest and more imaginative contributions were being made. These developments offered encouragement that management science will be better able to support the needs of top level management.

Gerald Hoffmann pointed out that management science was an integrative discipline which draws its strength as much from the breath of its perspective as from the depth of its analytic capability. To achieve the widest possible breath of perspective a management scientist should have the knowledge of the world around us and an appreciation of the philosophical and aesthetic ideals that distinguish man from other species.

I agree with these remarks.

Mathematical modelling is often put forward as an attractive possibility to enable a manager to prescribe strategies for making an organization behave in some desired way. Modelling is certainly a powerful tool but there are many philosophical difficulties to be overcome before it can be applied with confidence in relevant management problems. One reason is imprecision. Whether one likes it or not, the world of the manager is a fuzzy environment. Any decision-maker who reasons out actions to manipulate the world around him deals in a world of some imprecision. This contention seems to suggest that models of management processes might more appropriately be based on fuzzy set technology than on the classical set technology that is now bcing used.

All the techniques that have been developed to aid the decision-maker assume that all the information is exact; that is, that there exists no special difficulty in the way of specifying the model's parameters. The principal complexity avoided by means of this kind of assumption concerns the need for some mechanism that will in a quantitative way, set in action in the minds of the assessors the merits of the alternatives, or to find common expressions of differences of opinions, as in group decision-making. In real management problems people normally cannot and do not make policy decisions by reducing their tradeoffs to a single number. The manager does not always perceive a scene with complete accuracy; when be does, it may still change.

The notions of enduring fuzziness and using fuzziness are enticing, and we believe that they may offer exciting new approaches to management studies. By enduring fuzziness we mean the capacity to tolerate inexact data in classical decision methods; in other words, to reduce the imprecision to a level of relative unimportance. By using fuzziness we mean the capacity to handle inexact data in order to get proficiency in solving real managerial situations, i.e. to scan the external environment.

7.1. OPTIMIZATION IN A FUZZY ENVIRONMENT

Work done over the past few years has fully demonstrated the practicability of using System Theory and fuzzy sets theory in real management problems. Attention is concentrated in this section on such problems, with special emphasis on planning methods. Fuzzy programming was found to be a useful tool either to tolerate ambiguity or to use it as a blessing. The fuzzy tolerance concept is introduced in order to represent beliefs, judgement and experience.

Since its birth at about the beginning of this century, the central task of management science has been to show how to make an organization behave in some desired way. In the simplest terms, an organization is an entity created to convert resources into finished products. Since this process is dynamic, it follows that its activity must be governed by a plan, i.e. an estimated evolution. Furthermore, this plan may alter with time and changing conditions. It is a major function of management to ensure that the plan is continually reviewed in order that it may be kept up to date as conditions change. System Theory is becoming an increasingly accepted framework for developing planning and control techniques. Consequently, planning and control will be considered to be trajectory estimations and trajectory maintenance; by trajectory we mean the evolution of a dynamic system, the production process.

Central to all this is the notion of a model, a model that involves a multistage structure. The solution of the model is a plan. The production policy is one of feedback, tending to make the actual system less sensitive to variations of parameters, and making use of the most up-to-date information. If a disturbance occurs, the error is processed, and a correction is applied to the initially planned values. A performance criterion is minimized to keep state and control variables near desired values. In this section attention is primarly concentrated on the planning phase in which the trajectory of a dynamic system is chosen and applied.

Any attempt to understand an industrial system requires us to abstract from the real situation relatively few variables of interest. According to modern System Theory any description of a system requires not only that we specify the inputs and outputs, but also that we have intervening variables that constitute the internal state. Economists are dealing routinely with the 'stock' concept. A stock represents a quantity fixed at a point in time. Inventories and capital equipment are thought of in this way. To build a dynamic model of an industrial process is to identify the stocks of unfinished production and their relationships. Practically, an industrial process can be modelled as a discrete-time system with a disturbance input due to raw materials supply and sales. The components of the state vector are backlogs x_i of parts awaiting to be processed. The control variables are the parts u_i scheduled for various work stations.

Let us take an example of the manufacturing of a device that consists of two parts. The operations u_1, u_2 process the raw materials x_1, x_3. The two resulting parts x_2, x_4 merge in the assembling operation u_3. The process under consideration consists of machining and assembly operations. Each of these requires a known amount of time. State variables represent the number of parts that are backlogged and waiting to be processed in the following points. β means how many parts must be subtracted from a backlog when two production paths merge in the assembling operation; α means specific consumption.

The state equation (SE) for the system can be written as:

$$\begin{aligned} x_1(k+1) &= x_1(k) - \alpha_1 u_1(k) + f_1(k) \\ x_2(k+1) &= x_2(k) + u_1(k) - \beta_2 u_3(k) \\ x_3(k+1) &= x_3(k) - \alpha_2 u_2(k) + f_2(k) \\ x_4(k+1) &= x_4(k) + u_2(k) - \beta_3 u_3(k) \\ x_5(k+1) &= x_5(k) + u_3(k) - f_3(k) \end{aligned} \tag{SE}$$

It is assumed that management has a long-term raw materials supply and sales rate available for planning purposes, i.e. that there exists a set of nominal values f_1, f_2, f_3. Therefore, the state equation describes the production flow at various moments and shows how much should be processed at specific moments to ensure the carrying out of certain quantities imposed by strategical planning. The underlying philosophy of the model is the following: a stock of today is equal to the stock of yesterday, to which is added what was produced from yesterday until today, and from which is subtracted what was delivered from yesterday. To these equations one must add constraints imposed by available machinery, and technological succession. If, for instance, u_1 and u_2 are operations executed by the same machine, then we must write

$$\gamma_1 u_1(k) + \gamma_2 u_2(k) < M$$

To execute operation u_1 is to satisfy the constraint

$$x_1(k) - \alpha_1 u_1(k) \geqslant 0$$

Many descriptions of production processes may finally, without too radical a distortion of the facts, be reduced to a mathematical formulation of precisely the kind suggested. A general formulation of many of these—of both strategical or operational type—possess the following structure

$$x(k+1) = Ax(k) + Bu(k) + Cf(k)$$
$$Px(k) + Qu(k) \geqslant R(k),\ x(k) \geqslant 0,\ u(k) \geqslant 0$$

where A, B, C, P, Q, R are constant matrices.

Very often the coefficients α, β have to be thought as fuzzy sets. A fuzzy set $\mathbb{R} \to [0, 1]$ can be given as a rule by an expert; an expert is considered here to be an individual recognized because of his training or experience as having considerable knowledge and understanding of a particular field. The assessment of the parameters α, β can be organized to bring out the merits of the alternatives in the minds of the assessors in a quantitative manner. The ability to work with fuzzy tolerances instead of numbers seems to be exactly the way people cope with their environment. The fuzzy version of SE (FSE) can be written as follows:

$$\begin{aligned} &x_1(k+1) - x_1(k) + \underset{\sim}{\alpha}_1 u(k) \subseteq \underset{\sim}{f}_1(k) \\ &x_2(k+1) - x_2(k) - \underset{\sim}{\beta}_2 u_3(k) \subseteq \underset{\sim}{0}(k) \\ &x_3(k+1) - x_3(k) + \underset{\sim}{\alpha}_2 u_2(k) \subseteq \underset{\sim}{f}_2(k) \\ &x_4(k+1) - x_4(k) - u_3(k) + \underset{\sim}{\beta}_4 u_3(k) \subseteq \underset{\sim}{0}(k) \\ &x_5(k+1) - x_5(k) - u_3(k) + \tilde{f}_3(k) \subseteq \underset{\sim}{0}(k) \end{aligned} \qquad \text{(FSE)}$$

where $\sim$ means fuzzy set $\mathbb{R} \to [0, 1]$, and $0(k)$ is also a fuzzy set having the form e^{-k}.

In this section robustness means the ability to work with models having fuzzy parameters. Thus, the robust programming problem for the linear case may be stated in the following manner: select values for a set of variables $x_1, x_2, \ldots, x_n$ so as to maximize an objective function $f(x_1, \ldots, x_n)$ subject to $\sum_{j=1}^{n} a_{ij}x_j \subset b_i$, where $a_{ij}: R \to [0, 1]$ and $b_i: R \to [0, 1]$ are fuzzy sets. To permit this formulation it is imperative that we can add two fuzzy sets and multiply a fuzzy set by a number. In order to speak about addition and multiplication of fuzzy sets, we recall at this stage that if $g: W \to Z$ is a mapping and a a fuzzy set, then the image of a through g is a fuzzy set $g(a): Z \to [0, 1]$, given by $g(a)(z) = \sup a(w)$ $w \in g^{-1}(z)$ if $g^{-1}(z) \neq \phi$, that is, where the supremum is taken over the set of points $g^{-1}(z)$ in W which are mapped by g into z.[51]

If the carrier W is a vector space, i.e. addition and the multiplication are defined by a scalar, these operations can be extended to the fuzzy sets of this vector space, by using the image of the intersection and of the assignment $w \mapsto kw$. The sum of two fuzzy sets will be defined therefore as

$$(a_1 + a_2)(z) = \sup\,[a_1(w_1) \wedge a_2(w_2)] \quad \text{for} \quad w_1 + w_2 = z$$

where $\wedge$ stands for inf and the product as $a(kw)$.

Example 1: Let $a = \left(\frac{2}{0.8}; \frac{3}{0.2}; \frac{4}{0.4}\right)$, $b = \left(\frac{4}{0.7}; \frac{5}{0.5}; \frac{6}{0.6}\right)$, $x = 5$

Then $a + b = \left(\frac{6}{0.7}; \frac{7}{0.5}; \frac{8}{0.6}; \frac{9}{0.4}; \frac{10}{0.4}\right)$

$$5a = \left(\frac{10}{0.8}; \frac{15}{0.2}; \frac{20}{0.4}\right)$$

For instance

$$(a + b)(8) = \max\{\min[a(2), b(6)], \min[a(3), b(5)] \\ \min[a(4), b(4)]\} = 0.6$$

Having brought matters to this point, we are ready to pass to the difficult matter of the solution at a level of detail adequate for computation. First, we make the important observation that, according to the representation theorem, any fuzzy set may be represented by a family of sets, called levels. A level of a fuzzy set a_i is a set

$$A_r^i = \{\alpha \in R \,|\, a_i(\alpha) > r, r \in [0, 1]$$

with the evident properties

$$r_1 \leqslant r_2 \Rightarrow A_{r2} \subset A_{r1}$$

$$a_i(\alpha) \leqslant a_j(\alpha) \Rightarrow A_r^i \subset A_r^j$$

We are now able to specify the feasible region via simple set constrainment

$$A_r^1 x_1 + A_r^2 x_2 + \ldots + A_r^n x_n \subset B_r$$

In this way a fuzzy optimization problem was replaced by an infinite number of inexact linear programming problems via an auxiliary linear programming problem $Wx \leqslant c$, since the resource set has the special form $B = \{y \in R^m | y \leqslant c\}$. To obtain the activity vectors w_j for the auxiliary linear programming problem, for each j we define the column w_j whose ith component is sup α_{ij}, $\alpha_j \in A_r^j$. The possibility of an infinite number of problems seems to preclude the possibility of solving the initial problem. In fact, when the fuzzy set is sampled we can manage it. Turning back to our FSE and sampling the unit interval as $\{0, \frac{1}{3}, \frac{2}{3}, 1\}$ the first equation is replaced by three equations

$0 \qquad x_1(k + 1) - x_1(k) + [\alpha_{1m}, \alpha_{1M}]u_1(k) \subseteq [f_{1m}, f_{1M}]$

$\frac{1}{3} \qquad x_1(k + 1) - x_1(k) + [\alpha'_{1m}, \alpha'_{1M}]u_1(k) \subseteq [f'_{1m}, f'_{1M}]$

$\frac{2}{3}$ $x_1(k+1) - x_1(k) + [\alpha''_{1m}, \alpha''_{1M}]u_1(k) \subseteq [f''_{1m}, f''_{1M}]$

We can convert the problem involving inclusion into one involving inequality constraints. For instance, the first equation becomes

$$x_1(k+1) - x_1(k) + \alpha_{1M}u_1(k) \leqslant f_{1M}$$

$$x_1(k+1) - x_1(k) + \alpha_{1m}u_1(k) \geqslant f_{1m}$$

The price we pay for this transformation is enlarging the dimensionality of the initial problem. The new variables and constraints are, however, exactly those we are used to handling.

Another way to approach this problem is to define the fuzzy tolerances f_1, f_2 as mappings $\mathbb{R} \to [0, 1]$ where

$$\delta_1 \in [f_{1m}, f_{1M}]$$

$$\delta_2 \in [f_{2m}, f_{2M}]$$

For instance $f_i(\delta) = \sin \pi\ (\delta - f_{im})/(f_{1M} - f_{im})$

Let $(x_1(k+1) - x_1(k) + \alpha_1 u_1(k) - f_{1m})/(f_{1M} - f_{1m}) = \varphi_1$

$(x_3(k+1) - x_3(k) + \alpha_2 u_2(k) - f_{2m})/(f_{2M} - f_{2m}) = \varphi_2$

We are able now to write two new objective functions $\sin \pi\ \varphi_1$, and $\sin \pi\ \varphi_2$, facing a fuzzy programming problem: Find $u_0, u_1, \ldots, u_{N-1}, x_1, x_2, \ldots, x$ such that min min $(\sin \pi\ \varphi_1, \sin \pi\ \varphi_2)$ = sup min $(\sin \pi\ \varphi_1, \sin \pi\ \varphi_2)$.

7.2. PROGRAMMING AND SCHEDULING

This section deals with dynamic linear programming and its importance in management information systems. It is followed by a discussion on the impact the new theory of decision in fuzzy environment has on such systems.

The concept of information systems has long been considered by those concerned with the use of the computer. An information system means a system for automatic data-processing. The extensive application of computers in the economy has, however, demanded consideration of a special type of information system, namely, the management information system. That is why, before considering anything else, we must accept a definition of what management is. In the following we shall propound a widely based definition by including the aspects of forecast, control, and organization, in which ‘to manage’ means to forsee, to control, and to organize. The concept is valid irrespective of the object managed, which may be a technological process, the production of a plant, or a macro-economic system. The important fact is that a management system must ensure the execution of tasks of management. As the volume of necessary data for exercising a task has become bigger, so more severe difficulties have developed, bringing with them, the need for mechanization, automation for simplifying control, or operations research for ensuring forecast and organization models. The advent of the computer allowed, in addition, a simultaneous treatment of several tasks, starting with an emphasis on control. Stripped to the more basic, common essentials, management information systems are now error processors. Based on the concept of control by feedback, the adaptation to a changing environment provides the powerful paradigm of strategy formulation in planning, i.e. finding the best ways of using resources in achieving goals and objectives. The speed of making decisions is vital in a world of change.

Just as in its evolution computing equipment covered several generations, management information systems have also gone through several generations. At the beginning, the information system achieved a recording of data, just as mechanical systems did, so that the decision-maker could see 'what was happening'. Then, it used algorithms by means of which static planning models were solved. The computer allowed a rapid handling of a big volume of data. When the volume increased, special problems arose in connection with the organization of those data, and this is how the concept of the database appeared. Later, it was claimed that the managed object should be considered as a whole, the concept of an integrated system was formulated, and the cybernetic approach of management appeared. Finally, management information systems have come under attack from all quarters as being inaccurate and inappropriate. The flexibility to function effectively in turbulent environments requires dynamic models to predict change and the effect of change. Central planning for massive resource allocation needs an extended time horizon. We shall dwell on this very fact.

First, let us see what classical linear programming offers; this technique is defined as an analysis of problems in which a linear function of several variables must be maximized or minimized when these variables are subjected to constraints, generally in the form of linear inequalities. Such problems occur when seeking an allocation of limited resources for achieving a desired objective, cost or profit. These problems are characterized by a number of solutions that satisfy the imposed constraints. The choice of a certain solution depends on the objective that is chosen. Any linear programming problem, specified as above, can be solved with a peak-searching technique that is processed by a computer in a parametric program deck.

A plan elaborated with such a model, however, does not take into account the evolution in time of the managed object. To include a time factor in the model a dynamic model may be used, one which describes the trajectory of the managed object, namely, the actual unfolding of the plan. To construct dynamic models one applies the concepts of state and control from the theory of systems: the construction of a dynamic model means to write state equations of the managed object.

Let us take the example of an industrial production process modelled as a discrete-time system with a disturbance input due to sales. The state consists of backlogs of parts awaiting processing. The control variables are the parts scheduled for various work stations. The underlying philosophy of the model is as follows. The production stock of today is equal to the stock of yesterday to which is added what was produced from yesterday until today, and from which is subtracted what was delivered yesterday. If all the production intermediate unfinished stocks for all jobs are taken into account, together with all their interrelationships, a series of equations is obtained which together represent the vectorial state equation of the production process. If to these equations are added constraints imposed by production capacity, by the succession of operations, and by deliveries, a production model is obtained. In this way, a new type of constraint is added to the classical linear programming problem, i.e. the dynamics of the managed object which describes production flow at various moments. The state equation describes this flow and shows how much should be produced at specific moments to ensure the completion of certain quantities imposed by strategical planning.

Many descriptions of decision problems may finally be reduced to a mathematical formulation of precisely the kind suggested without too radical a distortion of the facts. It is true that the initial formulation of many of these—both strategical and operational types—will possess the following structure:

$$x_{k+1} = Ax_k + Bu_k + Cf_k$$
$$Px_k + Qu_k < Rg_k$$
$$x_k \geqslant 0$$
$$u_k \geqslant 0$$

where x_k, u_k, f_k are vectors representing backlogs, scheduled parts, and deliveries, respectively. If we add an objective function the problem may be approached as a linear programming problem of considerable size because the variables as well as the constraints are obtained by multiplying the dimensions of vectors by the number of stages. The computer becomes less effective as the size of the problem increases. The more stages there are, the more necessary it is to find new approaches in order to reduce the operations at the stage level. Research in the field is trying to solve this problem.

The existence of dynamic models is useful not only for correctly formulating a forecast of a plan or programme but also for checking and control. In most management information systems checking means measuring indices and drawing up reports that show the results at a given moment and, eventually, deviations from the planned figures. It is natural to try (on the basis of these deviations) to obtain a clue about a correction decision, i.e. to annual this deviation. Something like that can be done with an algorithm, which should process the deviation.

Our overall position at the moment may briefly be recapitulated as following. For some given description of a planning problem we seek its mathematical formulation in terms of a model that involves a multi-stage structure whose associated mathematical problem takes the form of a dynamic linear programming; the solution of this problem is the plan. The production policy is a feedback policy, tending to make the managed object less sensitive to variations of parameters and making use of the most up-to-date information of the state of the production system. If a disturbance occurs, the error is processed and a correction is applied to the planned values. A quadratic performance criterion is minimized so as to keep state and control variables near desired values.

It is clear that even a simple and seemingly natural problem description may generate a host of nonroutine mathematical considerations.

Finally, the task of organization—which the classical information systems have so far not approached—can also be accomplished by means of a dynamic model. This time, the model is used for simulation, to predict change and the effect of change. In fact, the model is a representation of the connections existing between the elements of the managed object, and these connections reflect a certain existing pattern of organization. If we modify these connections there evolves another pattern of organization with a modified model. It is thus obvious that, through simulating various modes of organization by adequately modifying the model we can see 'what happens if'. The advantage is clear, for without changing the real system we can anticipate the effects of the changes. It should now be said that the use of a dynamic model is absolutely necessary for a developed information system with which to accomplish the task of organization. It allows trajectories of the system to be routed in various predictable situations, trajectories which, in most cases, are counter intuitive, i.e. not in accordance with expectations dictated by simple reasoning or commonsense. This is only natural, for production systems are complex and their development is dictated by a large number of factors,some of them powerfully influenced by the complex behaviour of people.

Management is currently reaching the responsive frontier where the new skills of ambiguity resolution will become more critical. Decisions in production systems are made by people and formulated in a natural language, which is strongly affected by ambiguity. Moreover, in contradistinction to technical systems, for accomplishing the task of checking the measuring of results is done by interrogating people; they again resort to natural language; and the lack of precision is a reality. Decision-making in uncertain situations is characteristic of man and it should be realistically considered in the design of any management system. To use deterministic models supplied with imprecise data or stochastic models, when we do not have the required statistics, leads to wrong solutions. A recent, and simple, technique that uses the latest results of fuzzy systems theory[112] seems to be of merit and the first applications appear to be promising. The fuzzy models seem to represent greater reality, and possess the advantage of relying on relatively simple mathematics much more suitable for computers. We shall now turn back to the fuzzy programming problem presented in Section 6.2. Take the case of planning in fuzzy environment. In many managerial situations to plan simply means to solve a dynamic system; for instance, if a production-inventory system is considered, some constraints regarding production capacities, deliveries, and stocks must be satisfied. Let the production-inventory system be

$$x(k+1) = d(x(k), u(k)) + f(k)$$
$$u(k) \in U_k(x(k))$$
$$x(k) \in X_k$$

where x means stock, u means production, and f means delivery. Given the sequence $f(k)$, $k = 0, 1, \ldots, N-1$, we can face the situation when there is no control sequence $u(k)$ to satisfy both dynamics and constraints. To face a problem that has no solution, a modification of constraints seems imperative. The planner is eager for an indication of the changes of constraints that will even yet help him to find a solution. In fact, he wants to know the minimum degradation needed in the specifications to make the problem solvable. It is evident that proficiency in changing variables f could be a way of finding a solution. But how far? We will suppose that our planner operates in a flexible environment, i.e. instead of numbers he handles intervals.

The problem is to find a certain $f(k)$ as close as possible to $\bar{f}(k)$. 'As close as possible' is modelled as a fuzzy set $f(k): R^n \to [0, 1]$ having the support $[\underline{f}(k), \bar{f}(k)]$. We remind the reader that the support of a fuzzy set is the set $\{z \in R^n \mid f(k)(z) > 0\}$.

With the above problem we associate the following problem:

$$\begin{aligned} \text{Find} \quad & \max \min f(k)\,[x(k+1) - d(x(k), u(k))] \\ & x(k), u(k), k = 0, 1, \ldots, N-1 \\ \text{subject to} \quad & u(k) \in U_k(x(k)) \\ & x(k) \in X_k \end{aligned}$$

Since $U_k(x(k)) \neq \phi$, $X_k \neq \phi$ by hypothesis, then this problem always has a solution. The problem is equivalent to:

$$\begin{aligned} \text{Find} \quad & \max r \qquad r \in R \\ \text{subject to} \quad & r \leqslant f(k)\,[x(k+1) - d(x(k), u(k))] \\ & k = 0, 1, \ldots, N-1 \\ & u(k) \in U_k(x(k)) \\ & x(k) \in X_k \end{aligned}$$

The inequality may be transformed into an equality by using a slack variable $s(k)$. Thus,

$$r + s(k) = f(k)\,[x(k+1) - d(x(k), u(k))]$$

If the fuzzy set $f(k)$, having the support $[\underline{f}(k), \bar{f}(k)]$ has the form

$$f(k)(t) = \frac{t - f(k)}{\underline{f}(k) - \bar{f}(k)} \qquad t \in R$$

then $$r + s(k) = \frac{x(k+1) - d(x(k), u(k)) - f(k)}{\underline{f}(k) - \bar{f}(k)}$$

and $$x(k+1) = d(x(k), u(k)) + \bar{f}(k) - (\underline{f}(k) - \bar{f}(k)(r + s(k))$$

It is easy to see that the control space has been enlarged, r and $s(k)$ being introduced as control inputs. The lack of precision can be assumed, for the sake of flexibility, or, in other words, the contradiction generating unsolvability can be avoided by compromise which generates possible alternatives.

We shall now turn our attention towards the scheduling problem. Scheduling may be defined as the assignment of work to a facility, and the specification of the time and the sequence in which the work is to be done. The time dimension of detail scheduling must be coordinated realistically with the master schedule of the end product. Detail scheduling is the process of breaking down the master schedule into departmental operations in order to synchronize the flow of work through several places. The main purpose in scheduling is to aid in setting up relative priorities among orders competing for capacity at a work centre. There is a widely held belief that scheduling requires accurate, detailed routings and time standards for all operations in order to make necessary calculations; this is patently ridiculous.[113] The task is to estimate the lead time for each operation to be scheduled. Lead time in management is made up of paperwork time, preparation (including setup time), processing time, move time, and queue (or delay time). To develop a schedule of orders moving through the plant, some estimates must be made of each of these elements. It is possible to calculate preparation and processing times and to make close estimates of paperwork and move times. However, in all but a very few real situations, the sum of all these elements of lead times will be no more that 10 per cent of the total. Queue times, delay times caused by backlogs of orders, account for the bulk of most lead times, and there is no practical way of making detailed estimates of this element. When this is recognized, it becomes obvious that precision is not the name of the scheduling game.

Loading to finite capacity assigns work to work centres in priority sequence so that the highest priority orders get the first claim to available capacity. When the limit of available capacity is reached, orders will have to be rescheduled earlier, if capacity is available, or dropped back into later periods until capacity to handle them is found. Obviously, this technique cannot be effective until a sound ordering system is available and until a computer program is operational to handle the masses of calculations involved.

Input control attempts to get work started on time, in the right order of priority, and with the greatest chance of finishing on time. However, many things interfere with the planned progress of work. The most effective criterion of priority is the due date of completion the order.

While it is possible to use days left until the due date as the priority ranking criterion, it is evident that some measure of days required to finish the work should also be considered. The ratio of this, too, is a common criterion of priority, and there are several others.

For example, consider two objects in the queue and two criteria. We may consider also the importance of each criterion, i.e. we can arrange the order of the two criteria. Sometimes we may wish to change the order in time, that is, we can on occasion emphasise the first criterion and at another emphasise the second. We have, therefore, two criteria $h_1, h_2 : R \to [0, 1]$, where R is the real line, and two functions $f_1, f_2 : R \to [0, 1]$. Now, let us fix the first object at time t and denote

$h_1(x) = a_1$ the priority of the first object according to the first criterion,

$h_2(y) = a_2$ the priority of the first object according to the second criterion.

$f_1(t) = b_1$ the importance of the first criterion at time t;

$f_2(t) = b_2$ the importance of the second criterion at time t.

Our intention is to determine a global priority taking into account all criteria. In order to do this we shall define a measure

$$g : P(A) \to [0, 1]$$

which is a set function defined on the monotone family of subsets of S, the set of priority values a_1, a_2 having the properties

$$g(\phi) = 0$$
$$g(A) = 1$$
$$\text{if } V, W \in P(A) \text{ and } V \subset W, \text{ then } g(V) \leqslant g(W).$$

For instance, we can have

$$g(a_1, a_2) = 1$$

$$g(a_1) = \sqrt{\frac{b_1^2}{b_1^2 + b_2^2}}$$

$$g(a_2) = \sqrt{\frac{b_2^2}{b_1^2 + b_2^2}}$$

$$g(\varphi) = 0$$

Now, in order to define the priority of an object according to the given criteria h_1, h_2, we shall introduce a function

$$p:\{h_1, h_2\} \rightarrow [0, 1]$$
$$p(h_1) = a_1$$
$$p(h_2) = a_2$$

If the numbers a_1, a_2 are ordered, in the sense that $a_1 < a_2$ the global priority is given by

$$[p(h_1) \wedge g(a_1, a_2)] \vee [p(h_2) \wedge g(a_2)]$$

which is precisely the fuzzy integral introduced by Sugeno.[114] This process is repeated for the other three objects and, finally, four numbers are obtained. These numbers are then ordered, and this order represents the global view which synthesizes all the initial criteria.

7.3. INFORMATION RETRIEVAL

The object of this section is to introduce a set of models that are, or promise to be, of general value in an information retrieval theory.[115]

The study of fuzziness is essential to a proper understanding or man-to-machine and machine-to-man communication. Information retrieval systems call for a need to manage fuzziness and not merely to react to it.

Here we shall investigate the possibility of using an index as a model and basis for getting to understand how information that has been acquired and stored in the process of learning can be retrieved at will. The manner in which an index is compiled and used, and the retrieval process are similar. In retrieval we need language for description and a strategy for searching. In indexing, the body of the book, the 'contents', is replaced by simple descriptions which serve as a basis, or as guides, in helping us to decide which topics of information we want to retrieve; and the skill in composing and making use of the descriptions plays a powerful role in determining the effectiveness of retrieval. In consequence therefore, we are concerned with syntax, i.e. a set or rules for combining words in a sequence that conveys meaning. We call these combinations, which extend the descriptive capacity of a vocabulary, 'coalitions' and use a fuzzy measure to describe their importance.

The principal objective of this section will be the formal statement of what we mean by a descriptors assignment problem.

Since the search criteria are based on the content of an information object, it becomes necessary to use content identifiers, such as a set of descriptors attached to each object normally chosen from a controlled list of allowable terms. The preparation of these sets is normally referred to as indexing.

Many different approaches to the problem of providing models for indexing have been explored by computer technologists, and a few have been tested by actual IR systems. It would be tedious and for our purposes unrewarding to attempt to describe individually even the most successful of these. Instead, let us discuss the general approach to the problem, which can be recognized in practically every system thus far reported.

Let us consider a set X of objects, a set Y of descriptors and a map $f: X \to \mathscr{P}(Y)$, where $f(x)$ is the subset of these descriptors assigned to the object x. We can consider also $g: B \to \mathscr{P}(x)$, where $g(y)$ is the subset of those objects having the descriptor y. It is obvious that f and g are, in fact, applications obtained from a relation defined on the Cartesian product $X \times Y$, namely, the relation 'the object x has the descriptor y'. Generally, the sets X, Y are considered to be finite. In the case, the relationship between objects and descriptors may conveniently be represented in the form of a matrix. For instance, the columns can represent a vocabulary of descriptors and the rows can represent objects indexed into a retrieval system. To enter an object into the system we assign it to classes by tagging it with appropriate values; this assignment is indicated in each cell of the matrix. Thus, the indexing process is the representation of objects by subsets of Y.

So far we have considered in our model sets and relations. We are ready now to note that an assignment is an arrow in category Rel, whose objects are sets and whose forms are relations between sets. The model of a decision-making task can, therefore, be studied in the framework of this explicit inclusion. If we note by id_x the identity morphism $X \to X: x \mapsto x$, we have got an abstraction of the identity relationship.

Identity is always understood as a binary relationship between a certain set of objects. The content of this relationship depends on the circumstances in which we are considering these objects, or on the observer who passes judgement on the identity of objects from his chosen point of view; identity turns out to be synonymous with interchangeability.

If the identity of objects signifies their complete interchangeability, then their resemblance means their partial interchangeability only. Such a relationship is given, for instance, by the coincidence of certain descriptors. We say, 'x_1, x_2 are partially indistinguishable if they have at least a common descriptor', and write $x_1 R x_2 \Leftrightarrow f(x_1) \cap f(x_2) \neq \phi$, if for each $x \in X$, $f(x) \neq \phi$; the relationship R is called a *tolerance*, i.e. one that is reflexive and symmetric. Indeed, every object resembles itself and two objects resemble each other, independently of the order in which we consider them, and transposition is by no means obligatory.

The decision-making model, however, is not well suited for dealing with real systems because it fails to come to grips with the reality of the fuzziness of descriptor assignment. Thus, we shall consider this fact in greater detail. We begin with a fundamental assumption concerning the existence of decision situations, the essential nature of which may conceivably be captured within the framework of a new model. Let us consider again a set X of objects, a set Y of descriptors and a map. This time the map will be $f: X \rightarrow \mathscr{F}(Y)$ or $g: Y \rightarrow \mathscr{F}(X)$, where $\mathscr{F}(X)$ are all the fuzzy subsets of X, $f(x)(y)$ is the grade of membership of the descriptor y to the object x, and $g(y)(x)$ is the grade of membership of the object x to the descriptor y. It is evident that f and g are, in fact, applications obtained from a fuzzy relation defined on the Cartesian product $X \times Y$. Generally, the sets X and Y are considered to be finite. Once again, the relationship between objects and descriptors may conveniently be represented in the form of a matrix. To enter an object into the system we assign it to fuzzy sets, by tagging it with membership degrees. In this context, to index is to represent objects by fuzzy subsets of the set Y.

We have considered in our new model sets and fuzzy relations. It is precisely for this reason that a fuzzy assignment can be viewed as an arrow in the category FRel, whose objects are sets and whose structures are fuzzy relations between sets. The strength of the model for fuzzy assignment lies in the fact that we can largely exploit the resemblance relation. From the beginning we shall stress that the condition of reflexivity can be replaced by a weaker one. A fuzzy resemblance relation may be defined as a mapping $t: X \times X \rightarrow [0, 1]$, such that $t(x, x) > 0$ and $t(x, y) = t(y, x)$.

To illustrate this idea in a concrete setting it may be well to consider a series of examples in which a fuzzy resemblance is given in various ways. For instance, bearing in mind the idea that two objects have at least a common descriptor in order to resemble each other, we can introduce the relationship $t(x_1, x_2) = \bigvee_{y \in Y}[f(x_1) \wedge f(x_2)]$.

Let us, now, examine the case where two objects have as many common descriptors as possible. Even though we consider the deterministic model X, $Y, f: X \to \mathscr{P}(Y)$, if we denote by card: $\mathscr{P}(Y) \to \mathbb{N}$ the map assigning to every part of Y the number of its elements, and by $b: \mathbb{R}_+ \to [0, 1]$ the fuzzy set of big numbers, then

$$h(x_1, x_2) = b\{\text{card}\,[f(x_1) \cap f(x_2)]\}$$

is a fuzzy relation on X. In the same way, if X, $Y, f: X \to \mathscr{F}(Y)$ is a fuzzy assignment model and $b': \mathscr{F}(Y) \to [0, 1]$ a fuzzy subset, then $h(x_1, x_2) = b'[f(x_1) \cap f(x_2)$ is a fuzzy resemblance relation.

Our overall position at this moment may be briefly recapitulated as follows: the description of a decision-making assignment is the mapping

$$F: X \times Y \to \{0, 1\}$$

with the property $\forall x \in X, \exists y \in Y$ with $F(x, y) = 1$, meaning that all the objects can be characterized by Y. F is a relation and $F(x, y) = 1$ means 'x has the descriptor y'. A descriptor assignment induces a mapping

$$f: X \to \mathscr{P}(Y)$$

defined as $f(x) = \{y \in Y | F \subset x, y) = 1\}$. Clearly, $\phi \in f(x)$. We denote by (F, X, Y) the model of descriptor assignment. The model (F, X, Y) induces a set of relations, as follows:

(*a*) 'two objects $a, b \in X$ have at least a common descriptor', defined δRb if $f(a) \cap f(b) \neq \phi$
Clearly, this relation is a tolerance.

(*b*) $X \times X \to \mathbb{N}$ defined as $(a, b) \to \text{card}\,[f(a) \cap f(b)]$ the number of common descriptors.

(*c*) 'as many common descriptors as possible' defined as
$R(a, b) = H[\text{card}\,(f(a) \cap f(b))]$

A fuzzy descriptor assignment is a mapping

$$F: X \times Y \to [0, 1]$$

with the property $\forall x \in X, \exists y \in Y$ with $F(x, y) > 0$. Therefore, F is a fuzzy relation. A fuzzy descriptor assignment induces $f: X \to \mathscr{F}(Y)$ and $F(x, y)$ means the membership degree of descriptor y to object x. The mapping $f: X \to \mathscr{F}(Y)$ is defined as $f(x)(y) = F(x, y)$. Clearly, $\phi \notin f(x)$. We denote by $(\tilde{F}, X, Y)$ a model of fuzzy descriptor assignment. Such a model induces the relationship 'two objects have at least a common descriptor' given by

$$p(a, b) = \bigvee_{y \in Y} [f(a) \wedge f(b)]$$

with properties

$$p(a, a) > 0$$

$$p(a, b) = p(b, a)$$

· We shall now consider, a cluster as a number of similar objects. This means that members within each cluster are sufficiently alike to justify ignoring the individual differences between them.

Let us consider a fuzzy measure $\mu : \mathscr{P}(Y) \to [0, 1]$ defined on Y. Then, a classification

$$\delta : X \to [0, 1]$$

is a global evaluation procedure defined as an assignment

$$a \to \int_{\mathscr{F}} f(a) \circ \mu$$

where $\int$ is the fuzzy integral as defined by Sugeno. [114,116] Note that fuzzy measures also have other definitions.[83,117]

Information is retrieved in response to a request. An IR system is, therefore, defined as a purposeful, and then subjective recovery. We shall now consider the problems of formulating a statement of information needs and the use of this statement to search the desired request to be a word in a language. Sets of descriptors provide great flexibility in describing subjects. Because there is no need of a structural relationship between descriptors, they can be added to or deleted from the vocabulary at will, making the language highly adaptable to subject matter changes.

A request is the basis for an evaluation by computing a relationship (identity; resemblance). Using a resemblance relationship the IR system is, then, one that compares the request with the description of stored objects and arranges all the objects in degree.

Let us consider the model (F, X, Y). We shall define $X_1 = X \cup \{q\}$, where q is the request. In this way a new model (F_1, x_1, Y) is constructed, with $F_1/x \times y = F$. The problem now is to find $a \in X$ as close to y as possible.

Let us consider the classification $\delta : X_1 \to [0, 1]$. The problem is solved by maximizing it.

$$d[\delta(a), \delta(q)]$$

It should be observed at this point that ambiguity and consequent retrieval of irrelevant objects can arise by ignoring relational information.

Various descriptors modify one another, greatly enlarging the number of individual statements that can be made. This is well illustrated by the example of the ambiguity between blind Venetians and Venetian blinds which occurs through failing to note when one descriptor is used to qualify another; the human brain takes advantage of the context; the coalition is used to avoid ambiguity. Almost any common word can lose its straightforward character by finding itself in company that brings out some hidden weakness. The human brain, guided by experience, has developed a capacity to cope with problems of this type.

The first step in our model will then be to define the grade of importance of every coalition and to investigate the consequences of this line of thinking. A possible way of evaluating this importance is to use the fuzzy measure.

Let us consider a set X of objects, a set Y of descriptors, and a map $f: X \rightarrow \mathscr{F}(Y)$. This is our model of fuzzy assignment (indexing), where $f(x)(y)$ is the link between the object x and the property y. The next step is to consider a subjective measure $\mu: \mathscr{F}(Y) \rightarrow [0, 1]$, expressing the grade of importance of a subset of descriptors. In this way we have introduced a tool for subjective evaluation of the context. The μ measure grades the groups of descriptors according to syntax.

Bearing in mind that $f(x) \in \mathscr{F}(Y)$, we are now ready to introduce the concept of syntax estimation as a function, with monotonicity defined by using fuzzy measures, i.e. the number $F_x = \int f(x)(y) \circ \mu = \sup [\alpha \wedge \mu(Y \cap X_u)]$ where $X_u = \{x | f(x)(y) \geqslant {}_{\alpha}\}$. In this way we have a new application

$$F: X \rightarrow [0, 1],\ F(x) = F_x$$

induced by a global criterion, the μ measure.

Let us consider, again, a set X of information objects and a set Y of descriptors. We can make statements about an element $x \in X$ of the form: the result of performing a test $y \in Y$ on $x \in X$ is $v \in V$. The basic predicates, out of which the descriptions of information objects are built, consist of primitive statements of the form $y(x) = v$. The statement

$$y_n(x_i) = v_1 \wedge \ldots \wedge y_n(x_i) = v_n$$

may be considered to describe the object x_i.

So far we have discussed two cases, $v = \{0, 1\}$ and $V = [0, 1]$. Now, we shall focus our attention on another case, when V is the set of fuzzy sets. For instance, the statement 'colour (x) = red' is a primitive statement, where 'colour' is a descriptor and 'red' is a value.

Then, the vocabulary can be thought as the set of values, and a keyword is a family of values. We can consider the descriptors $y \in Y$ as fuzzy sets $f_j : S_j \to [0, 1]$, where the key word S_j is a particular subset of descriptors $S_j \subset Y$.

For instance:

field = (mathematics, chemistry, physics, art)
language = (English, French, Russian, German)
year = (1900, 1901 . . . .)
pages = (100, 120 . . . .)

$f_3 : S_3 \to [0, 1]$ means new
$f_4 : S_4 \to [0, 1]$ means big

Let μ be a fuzzy measure on $\{1, 2, 3, 4\}$.
A query is a vector (f_1, f_2, f_3, f_4).

The answer of the system is an $x \in X$ which maximizes

$$\int f(x) \circ \mu$$

where f is a function resulting from $f_1, \ldots, f_4$.

So far it has been supposed that the user has a detailed knowledge of the vocabulary. Unfortunately, this is not the case. The process of request formulation is complex and depends on particular attributes of the requestor, such as his knowledge of the content of the store, his familiarity with the topic matter being searched, his personal preferences for vocabulary and style, and so on.

This is the reason that a relevance feedback is used to modify the original request within a learning process. Relevance feedback means that the requestor asks a question, gets his response, and retreats to ponder the output. He then has the opportunity to return and ask a new question.

REFERENCES

1. S. Klaczko, 'Pattern Recognition as a Model for Cognitive Processes'. In H. Bossell, S. Klaczko & N. Muller (eds.) *Systems Theory in the Social Sciences* (Birkhäuser Verlag, Basel, 1976).
2. H.M. Hersh and A. Caramazza, 'A Fuzzy Set Approach to Modifiers and Vagueness in Natural Language' *J. Exp. Psychol: Gen.*, **105**, 254–276 (1976).
3. N. Georgescu-Roegen, *The Entropy Law and the Economic Process* (Harvard University Press, Cambridge, Mass., 1971).
4. M.A. Aizerman, 'Some Unsolved Problems in the Theory of Automatic Control and Fuzzy Proofs' *IEEE Transactions on Automatic Control* **AC-24**, 146–148 (1977).
5. B.R. Caines and L.J. Kohout, 'The Fuzzy Decade: A Bibliography of Fuzzy Systems and Closely Related Topics' *Int. J. Man-Machine Studies* **9**, 1–68 (1977).
6. L.A. Zadeh, 'Fuzzy Sets' *Information and Control* **8**, 338–353 (1965).
7. L.A. Zadeh, 'Fuzzy Sets' in J. Belzer, A. Holzman and A. Kent (eds.) *Encyclopedia of Computer Science and Technology* (M. Dekker, New York, 1977).
8. A. Kaufmann, *Introduction to the Theory of Fuzzy Subsets* (Academic Press, New York, 1975).
9. L.A. Zadeh, K.S. Fu, K. Tanaka and M. Shimura (eds.) *Fuzzy Sets and their Applications to Cognition and Decision Processes* (Academic Press, New York, 1975).
10. C.V. Negoita and D.A. Ralescu, *Applications of Fuzzy Sets to System Analysis* (Birkhäuser Verlag, Basel, 1975).
11. J.P. Boyd, 'Topoi as Models of Fuzzy Sets' *Paper Presented at the 22nd Annual Meeting of the Society for General Systems Research, Washington* (February, 1978).
12. M. Arbib, 'Book Review Essay' *Bull. Amer. Math. Soc.* **81**, 947–951 (1977).
13. Ch. Carlsson, 'NOAK 77. A Nordic Operations Research Congress' *Kybernetes* **7**, 164–166 (1978).
14. M. Gupta, 'IFAC Report: Round Table Discussion on the Estimation and Control in Fuzzy Environments' *Automatica* **11**, 209–212 (1975).
15. E.H. Mamdani and M. Gupta, 'The Second IFAC Round Table Discussion on

Fuzzy Automata and Decision Processes' *Proceedings of the IFAC 6th World Congress, Boston* (August, 1975).
16. H.J. Zimmermann, 'Fuzzy Mathematics. Theory and Applications' in J. Rose and C. Bilciu (eds.) *Modern Trends in Cybernetics and Systems* (*Proceedings of the Third International Congress of Cybernetics and Systems, Bucharest, August, 1975*) (Springer Verlag, Berlin, 1977).
17. L.A. Zadeh, 'Outline of a New Approach to the Analysis of Complex Systems and Decision Processes' *IEEE Transactions on Systems, Man and Cybernetics* **SMC-3**, 28–44 (1973).
18. F.E. Wenstøp, 'Deductive Verbal Models of Organizations' *Int. Man-Machine Studies* **8**, 293–311 (1976).
19. F.E. Wenstøp, 'Fuzzy Set Simulations Models in a Systems Dynamics Perspective' *Kybernetes* **6**, 209–218 (1977).
20. F.E. Wenstøp, 'Verbal Formulations of Fuzzy Dynamic Systems' in J. Rose (ed.) *Current Topics in Cybernetics and Systems* (*Proceedings of the Fourth International Congress of Cybernetics and Systems, Amsterdam* August, 1978) (W.O.G.S.C. and Springer Verlag, Berlin, 1978).
21. E.H. Mamdani, 'Advances in the Linguistic Synthesis of Fuzzy Controlers' *Int. J. Man-Machine Studies* **8**, 669–679 (1976).
22. W.J. Kickert, 'An Example of Linguistic Modelling' *Report No. 30 Dept. of Ind. Engineering, University of Technology, Eindhoven*, The Netherlands (1976).
23. R.M. Tong, 'Synthesis of Fuzzy Models for Industrial Processes. Some Recent Results' *Int. J. Gen. Systems* **4**, 143–162 (1978).
24. W.J. Kickert and H.R. Van Nauta Lemke, 'Application of a Fuzzy Controller in a Water Plant' *Automatica* **12**, 301–308 (1976).
25. R. Jain, 'Outline of an Approach for the Analysis on Fuzzy Systems' *J. Control* **23**, 627–640 (1976).
26. P.S. King and E.H. Mamdani, 'The Application of Fuzzy Control Systems to Industrial Processes' *Automatica* **13**, 235–242 (1977).
27. C.P. Pappis and E.H. Mamdani, 'A Fuzzy Logic Controller for Traffic Junction' *IEEE Transaction on Systems, Man and Cybernetics* **SMC-7**, 707–717 (1977).
28. D.A. Rutherford and S.C. Bloore, 'The Implementation of Fuzzy Algorithms for Control' *IEEE Proceedings* **64**, 572–573 (1976).
29. R.M. Tong, 'Analysis of Fuzzy Control Algorithms Using the Relation Matrix' *Int. J. Man-Machine Studies* **8**, 679–686 (1976).
30. E. Sanchez 'Resolution of Eigen Fuzzy Sets Equations' *Int. J. Fuzzy Sets Systems* **1**, 69–74 (1978).
31. W.J. Kickert, 'Towards an Analysis of Linguistic Modelling' in *Current Topics in Cybernetics and Systems.* (*Proceedings of the Fourth International Congress on Cybernetics and Systems, Amsterdam, August 1978*) (W.O.G.S.C. and Springer Verlag, Berlin, 1978).
32. E. Sanchez, 'Resolution of Composite Fuzzy Relation Equations' *Information and Control* **30**, 38–48 (1976).
33. R. Lefaivre, 'The Representation of Fuzzy Knowledge' *Dept. of Computer Science, Rutgers University, New Brunswick* (January, 1974).
34. L.A. Zadeh, 'PRUF—a Meaning Representation Language for Natural Languages' *Memorandum ERL-M77/31, University of California, Berkeley* (Oct.,

1977).

35. J.M. Adamo, 'Implementation de la théorie des sous-ensembles floues. Application à l'analyse de processue de décision' *Thèse d'état. Université Claude Bernard*, Lyon (Mars, 1978).
36. J.M. Adamo, 'Application of Fuzzy Logic to the Design of a Bevioral Model in an Industrial Environment: An Application Case of the I.P.L. Language' in J. Rose (ed.) *Current Topics in Cybernetics and Systems* (*Proceedings of the Fourth International Congress of Cybernetics and Systems, Amsterdam, August, 1978*) (W.O.G.S.C. and Springer Verlag, Berlin 1978).
37. R. Bellman and L.A. Zadeh, 'Decision-Making in Fuzzy Environment' *Management Science* **17**, B, 141–164 (1970).
38. K. Asai and H. Tanaka, 'On the Fuzzy Mathematical Programming' *Proceedings of the 3rd IFAC Symposium on Identification and System Parameter Estimation, the Hague, June 1973.* (North-Holland, Amsterdam, 1973).
39. A.O. Esogbue, 'On the Application of Fuzzy Allocation Theory to the Modelling of Cancer Research Appropriation Process' in J. Rose and C. Bilciu (eds.) *Modern Trends in Cybernetics and Systems* (*Proceedings of the Third International Congress of Cybernetics and Systems, Bucharest, August, 1975*) (W.O.G.S.C. and Springer Verlag, Berlin, 1977).
40. C.V. Negoita, 'Fuzzines in Management' *Paper presented at the ORSA/TIMS Meeting, Miami* (November, 1976).
41. C.V. Negoita, 'On Dynamics and Fuzziness in Management Information Systems' in J. Rose and C. Bilciu (eds.) *Modern Trends in Cybernetics and Systems* (*Proceedings of the Third International Congress of Cybernetics and Systems, Bucharest, August, 1975*) (W.O.G.S.C. and Springer Verlag, Berlin, 1977).
42. C.V. Negoita, *Management Applications of System Theory* (Birkhäuser Verlag, Basel, 1979).
43. C.V. Negoita and D.A. Ralescu, 'On Fuzzy Optimization' *Kybernetes* **6**, 193–195 (1977).
44. C.V. Negoita and M. Sularia, 'On Fuzzy Mathematical Programming and Tolerances in Planning' *Economic Computation and Economic Cybernetics and Research* **1**, 3–14 (1976).
45. S.A. Orlovsky, 'On Programming with Fuzzy Constraint Sets' *Kybernetes* **6**, 197–202 (1977).
46. M. Sularia 'On Fuzzy Programming in Planning' *Kybernetes* **6**, 229–230 (1977).
47. H.J. Zimmermann, 'Optimization in Fuzzy Environment' *XXI Intern. TIMS and 46 ORSA Conf. Puerto Rico* (1974).
48. H.J. Zimmermann, 'Fuzzy Programming and Linear Programming with Several Objective Functions' *Int. J. Fuzzy Sets Systems* **1**, 45–56 (1978).
49. C.V. Negoita and D.A. Ralescu, 'Representation Theorems for Fuzzy Concepts' *Kybernetes* **4**, 169–174 (1975).
50. C.V. Negoita, 'Fuzzy Sets, System Theory and Management' in E. Billeter, M. Cuenod and S. Klaczko (eds.) *Overlapping Tendencies in Operations Research, System Theory and Cybernetics* (Birkhäuser Verlag, Basel, 1976).
51. C.V. Negoita, P. Flondor and M. Sularia, 'On Fuzzy Environment in Optimization Problems' in J. Rose and C. Bilciu (eds.) *Modern Trends in Cybernetics and Systems* (*Proceedings of the Third International Congress of Cybernetics and*

Systems, Bucharest, August, 1975) (W.O.G.S.C. and Springer Verlag, Berlin 1977).

52. C.V. Negoita, et al, 'On Considering Imprecision in Dynamic Linear Programming' *Econ. Comput. Econ. Cyb. Studies and Res.* **3**, 83–95 (1976).
53. C.V. Negoita, 'On Fuzzy Programming in Planning' in J. Janssen, L.F. Pau and A. Straszak (eds.) *Proceedings of the IFAC/IFORS/IIAS Conference on Dynamics Modelling and Control of National Economies, Vienna, January 1977* (North Holland, Amsterdam, 1979).
54. D. Dubois and H. Prade 'Systems of Linear Fuzzy Constraints' *Int. J. Fuzzy Sets Systems* **3**, 37–48 (1980).
55. C.V. Negoita, 'A Seminar on Economic System Dynamics' *Economic Computation and Economic Cybernetics Studies and Research* **3**, 109–110 (1976); also in *International Cybernetics Newsletter* **1**, **5** (1976) (*J. Cybernetics* **6**, 1–2, 1976).
56. C.V. Negoita and D.A. Ralescu, 'Some Results in Fuzzy System Theory' in J. Rose and C. Bilciu (eds.) *Modern Trends in Cybernetics and Systems* (*Proceedings of the Third International Congress of Cybernetics and Systems, Bucharest, August, 1975*) (W.O.G.S.C. and Springer Verlag, Berlin, 1977).
57. C.V. Negoita and Al. Stefanescu, 'Fuzzy Objects in Topoi: A Generalization of Fuzzy Sets' *Bull. Polytechnic Inst. Jassy* **28**, 7–9 (1978).
58. P. Flondor, 'C-Sets' *Communication at the Seminar of Fuzzy Systems, Faculty of Cybernetics, Bucharest* (1976).
59. D.A. Ralescu, 'Fuzzy Subobjects in a Category and the Theory of C-Sets' *Int. J. Fuzzy Sets Systems* **1**, 193–203 (1978).
60. M. Kochen, *Principles of Information Retrieval* (Mellville, Los Angeles, 1974).
61. H.J. Schek, 'Tolerating Fuzziness in Keywords by Similarity Searches' *Kybernetes* **6**, 175–184 (1977).
62. C.V. Negoita, 'On the Notion of Relevance in Information Retrieval' *Kybernetes* **2**, 161–165 (1973).
63. V. Tahani, 'A Fuzzy Model of Document Retrieval Systems' *Inf. Proc. Manag.* **12**, 177–187 (1976).
64. T. Radecki, 'New Approach to the Problem of Information System Effectiveness Evaluation' *Information Processing & Management* **12**, 319–326 (1976).
65. C.V. Negoita, *Information Storage and Retrieval* (Editura Academiei, Bucharest, 1970).
66. C.V. Negoita, 'On the Search Strategies in Automatic Information Systems' *Proceedings of the 6th International Congress of Cybernetics, Namur* (Sept., 1970).
67. C.V. Negoita, 'Fuzzy Mathematics: A New Paradigm' *Paper Presented at the 22nd Annual Meeting of the Society for General Systems Research, Washington* (February, 1978).
68. C.V. Negoita, 'On the Stability of Fuzzy Systems' *Paper Presented at the 1978 International Conference on Cybernetics and Society, Tokio* (November, 1978).
69. C.V. Negoita, 'On Fuzzy Systems' in J. Rose (ed.) *Current Topics in Cybernetics and Systems.* (*Proceedings of the Fourth International Congress of Cybernetics and Systems, Amsterdam, August, 1978*) (W.O.G.S.C. and Springer Verlag, Berlin, 1978).
70. J.A. Goguen, 'The Logic of Inexact Concepts' *Synthèse 19*, 325–373 (1968).
71. M.D. Mesarovic, D. Macko and Y. Takahara, *Theory of Hierarchical Multilevel*

Systems (Academic Press, New York, 1970).

72. M. Zeleny, 'The Theory of the Displaced Ideal' in *Multiple Criteria Decision Making* (Springer, Berlin, 1976).
73. J. Galbraith, '*Designing Complex Organizations*' (Addison-Wesley New York, 1973).
74. J.A. Goguen, 'Mathematical Representation of Hierarchically Organized Systems' in E.O. Attinger (ed.) *Global Systems Dynamics* (S. Krager, Basel, 1970).
75. M.A. Pollatschek, 'Hierarchical Systems and Fuzzy Set Theory' *Kybernetes* **6**, 147–152 (1977).
76. M. Mizumoto and K. Tanaka, 'Some Properties of Fuzzy Sets of Type 2' *Informations and Control* **31**, 312–340 (1976).
77. L.A. Zadeh, 'The Concept of Linguistic Variable and its Application to Approximate Reasoning' Part I, *Information Sci.* **8**, 199–249 (1975); Part II, *Information Sci.* **8**, 301–357 (1975); Part III, *Information Sci.* **9**, 43–80 (1976).
78. E.W. Chapin, 'Set-Valued Set Theory' *Notre Dame J. Formal Logic* **15**, 619–634 (1974).
79. L.A. Zadeh, 'A Fuzzy Interpretation of Linguistic Hedges' *J. Cybernetics 2*, 4–34 (1973).
80. M. Mizumoto, 'Fuzzy Sets under Various Operations' in J. Rose (ed.) *Current Topics in Cybernetics and Systems* (*Proceedings of the Fourth International Comgress of Cybernetics and Systems, Amsterdam, August, 1978*) (W.O.G.S.C. and Springer Verlag, Berlin, 1978).
81. A. DeLuca and S. Termini, 'On the Convergence of Entropy Measures of a Fuzzy Set' *Kybernetes* **6**, 219–227 (1977).
82. A DeLuca and S. Termini, 'Measures of Ambiguity in the Analysis of Complex Systems' in J. Gruska (ed.) *Mathematical Foundations of Computer Science* (Springer Verlag, Berlin, 1977).
83. H.T. Nguyen, 'On Fuzziness and Linguistic Probabilities' *J. Math. Anal. Appl.* **61**, 658–671 (1977).
84. A. Kandel and S.C. Lee, 'Fuzzy Switching and Automata Theory and Applications' (Crane and Russak, New York, 1979).
85. V.P. Srini, 'Realization of Fuzzy Forms' *IEEE Trans. Computers* **C-24**, 941–943 (1975).
86. Gr. Moisil, *Lectures on the Logic of Nuanced Reasoning* (Editura Stiintifica, Bucharest, 1975).
87. D. Ponasse, 'Algèbres floues et algèbres de Lukasiewicz' *Rev. Roumaine Math. Pure Appl.* No. 1 (1978).
88. E.D. Klement. 'Fuzzy δ-algebras and Fuzzy Meassurable Functions' *Institutbericht Nr. 104, Institut für Mathematik, Johannes Kepler Universität Linz* (February, 1978).
89. S. Gottwald, 'Universes of Fuzzy Sets Closed under Fuzzification' in J. Rose (ed.) *Current Topics in Cybernetics and Systems* (*Proceedings of the Fourth Congress of Cybernetics and Systems, Amsterdam, August, 1978*) (W.O.G.S.C. and Springer Verlag, Berlin, 1978).
90. P. Flondor, 'A Geometrical Sketch of Fuzzy Set Theory' *Communication at the Seminar of Fuzzy Systems, Faculty of Cybernetics, Bucharest* (1975).
91. D. Butnariu, 'L-Fuzzy Topologies' *Bull Soc. Sci. Math. de Roumanie* **19** (67),

227–236 (1975).

92. R. Lowen, 'Fuzzy Topological Spaces and Fuzzy Compactness' *J. Math. Anal. Appl.* **55**, 621–633 (1976).
93. H. Ludesher and E. Roventa, 'Sur les topologies floues definies *à* l'aide des voisinages' *C.R. Academie des Sciences Paris*, Serie **A.283**, 575–577 (Oct., 1976).
94. J. Michalek, 'Fuzzy Topologies' *Kybernetika 11*, 345–354 (1975).
95. S.S.L. Chang and L.A. Zadeh, 'Fuzzy Mapping and Control' *IEEE Transactions on System, Man and Cybernetics* **SMC-2**, 30–34 (1972).
96. S. MacLane, *Categories for the Working Mathematician* (Springer Verlag, Berlin, 1971).
97. M. Arbib and E. Manes, 'A Category-theoretic Approach to Systems in a Fuzzy World' *Synthèse* **30**, 381–406 (1975).
98. M. Arbib and E. Manes, '*Arrows, Structures and Functions: the Categorical Imperative*' (Academic Press, New York, 1975).
99. J.A. Goguen, 'Concept Representation in Natural and Artificial Languages: Axioms, Extensions and Applications for Fuzzy Sets' *Int. J. Man-Machine Studies* **6**, 513–661 (1974).
100. C.V. Negoita and Al. Stefanescu, 'On the State Equation of Fuzzy Systems' *Kybernetes* **4**, 213–214 (1975).
101. G.L. Shackle, *Decision, Order and Time in Human Affairs* (2nd ed. Cambridge, University Press, 1969).
102. C.V. Negoita, 'Fuzzy Models for Social Processes' in H. Bossel, S. Klaczko and H. Muller (eds.) *System Theory in the Social Science* (Birkhäuser Verlag, Basel, 1976).
103. W. Alston, 'Vagueness' in P. Eduards (ed.) *Encyclopedia of Phylosophy* (MacMillan, New York, 1967).
104. R.R. Verma. 'Vagueness and the Principle of Excluded Middle' *Mind* **79**, 67–77 (1970).
105. J.A. Goguen, 'L-Fuzzy Sets' *J. Math. Anal. Appl.* **18**, 145–174 (1967).
106. Y. Gentilhomme, 'Les ensembles floues en linguistique' *Cahiers de linguistique théorique et appliqué* **5**, 47–65 (1968).
107. M. Arbib and E. Manes, 'Fuzzy Machines in a Category', *Bull. Austral. Math. Soc.* **13**, 169–210 (1975).
108. G.C. Wraith, 'Lectures in Elementary Topoi' in F.W. Lawvere and A. Maurer (eds.) *Model Theory and Topoi* (Springer Verlag, Berlin, 1975).
109. S. MacLane, 'Sets, Topoi and Internal Logic in Categories' in H.E. Rose and J.C. Stepherdson (eds.) *Logical Colloquium* **'73** (*Proceedings of the Bristol Colloquium, July, 1972*) (North-Holland, Amsterdam, 1975).
110. S.S.L. Chang, 'Application of Fuzzy Set Theory to Economics' *Kybernetes* **6**, 203–207 (1977).
111. R. Rosen, 'Complexity as a System Property' *Int. J. Gen. Systems* **3**, 227–232 (1977).
112. R.M. Tong, 'A Control Engineering Review of Fuzzy Systems' *Automatica* **13**, 559–569 (1977).
113. W.G. Plossl, *Manufacturing Control. The Last Frontier for Profit* (Reston, Reston, Virginia, 1973).
114. M. Sugeno, 'Theory of Fuzzy Integrals and its Applications' *Ph.D. Thesis*,

Tokio Institute of Technology, Tokio (1974).
115. C.V. Negoita and P. Flondor, 'On Fuzziness in Information Retrieval' *Int. J. Man-Machine Studies* **8**, 711–716 (1976).
116. M. Sugeno and T. Terano, 'A Model of Learning Based on Fuzzy Information' *Kybernetes* **6**, 157–166 (1977).
117. E. Roventa, 'Fuzzy Measures' *Bull. Polytechnic Inst. Timisoara* **22** (36), 41–45 (1977).

INDEX

(including some names appearing only in the List of References, pages 102–108)